United States Field Artillery

Turner Publishing Company
Publishers of Military History
P.O. Box 3101
Paducah, Kentucky 42002-3101

Co-published by Turner Publishing Company
and Mark A. Thompson, Associate Publisher

Author: Philip A. St. John, Ph.D.
Graphic Designer: Elizabeth A. Dennis

Library of Congress Catalog
Card No. 96-61040
ISBN: 1-56311-313-9

Printed in the United States of America
Limited Edition

Front cover painting by artist Don Spaulding, courtesy of the Office of Chief Army Reserves.
Print provided by George J. Ellis

CONTENTS

Publishers's Note

We would like to thank those responsible for making this publication possible.

We owe our deepest gratitude to all the men and women of Field Artillery for giving of themselves, heart and soul and in some cases their lives, during their courageous service for our country.

We especially want to thank the U.S. Field Artillery Association, Anna Lou Johnson, Managing Director, General (Retired) Walter T. Kerwin, Jr., President Emeritus, and Lieutenant General (Retired) David E. Ott, President, for seeing this project through to its completion.

A special thanks goes to Philip A. St. John, Ph.D., for compiling information and writing the manuscript.

We are indebted to the many individuals who submitted photographs and additional material from which we got much of our information.

Turner Publishing Company leads the way in military association history book publishing, and we hope all will enjoy our newest title that chronicles the history of United States Field Artillery.

Dave Turner, President
Mark A. Thompson, Associate Publisher

David E. Ott
Lieutenant General, United States Army, Retired
President
United States Field Artillery Association

Walter T. Kerwin, Jr.
General, United States Army, Retired
President, Emeritus
United States Field Artillery Association

Foreword

Most of us have heard the axiomatic statement that those who don't understand history are doomed to repeat it. For this reason more than any other, the United States Field Artillery Association has taken on the task of publishing a comprehensive history of the King of Battle, detailing its pivotal role in every phase of the evolution of the profession of arms. In addition, this volume contains some of the lore, or "esprit," of the branch, as well as detailed history of the Association — knowledge useful to anyone wishing to preserve important traditions and move forward with the best interests of the Field Artillery at heart.

As a leading member of the combined-arms team, today's United States Field Artillery maneuvers with its fires using the most sophisticated command, control, communications, computers and intelligence (C^4I) infrastructure in the world. Yet, we have just begun to employ the rapid advances in technology that can now be applied to future combat systems development. Achievable imperatives well within our reach stand before us: improved weapons accuracy and lethality; increased range and deployability; and near-real-time situation awareness at all operation levels.

This work offers every member of the Field Artillery and combined-arms communities, and others interested in the military profession, an azimuth to the future that runs through the past. This history offers today's and tomorrow's leaders the knowledge to create change with continuity — a continuity that comes only from a thorough knowledge of the failures and triumphs of the past.

We hope all members and friends of this military family will make this valuable book a part of their personal libraries.

Gun crew of Battery C, 9th Battalion, 3rd Infantry Division, November 20, 1943. (Photo courtesy of G. William Mohr)

Men of gun crew #6 ready their 105-mm howitzer "Cong Fighter", 1968. (Photo courtesy of the National Archives)

A direct fire shoot, Chocolate Mountains aerial gunnery range in California, March 1994. (Photo courtesy of Brian E. Crane)

History of the United States Field Artillery

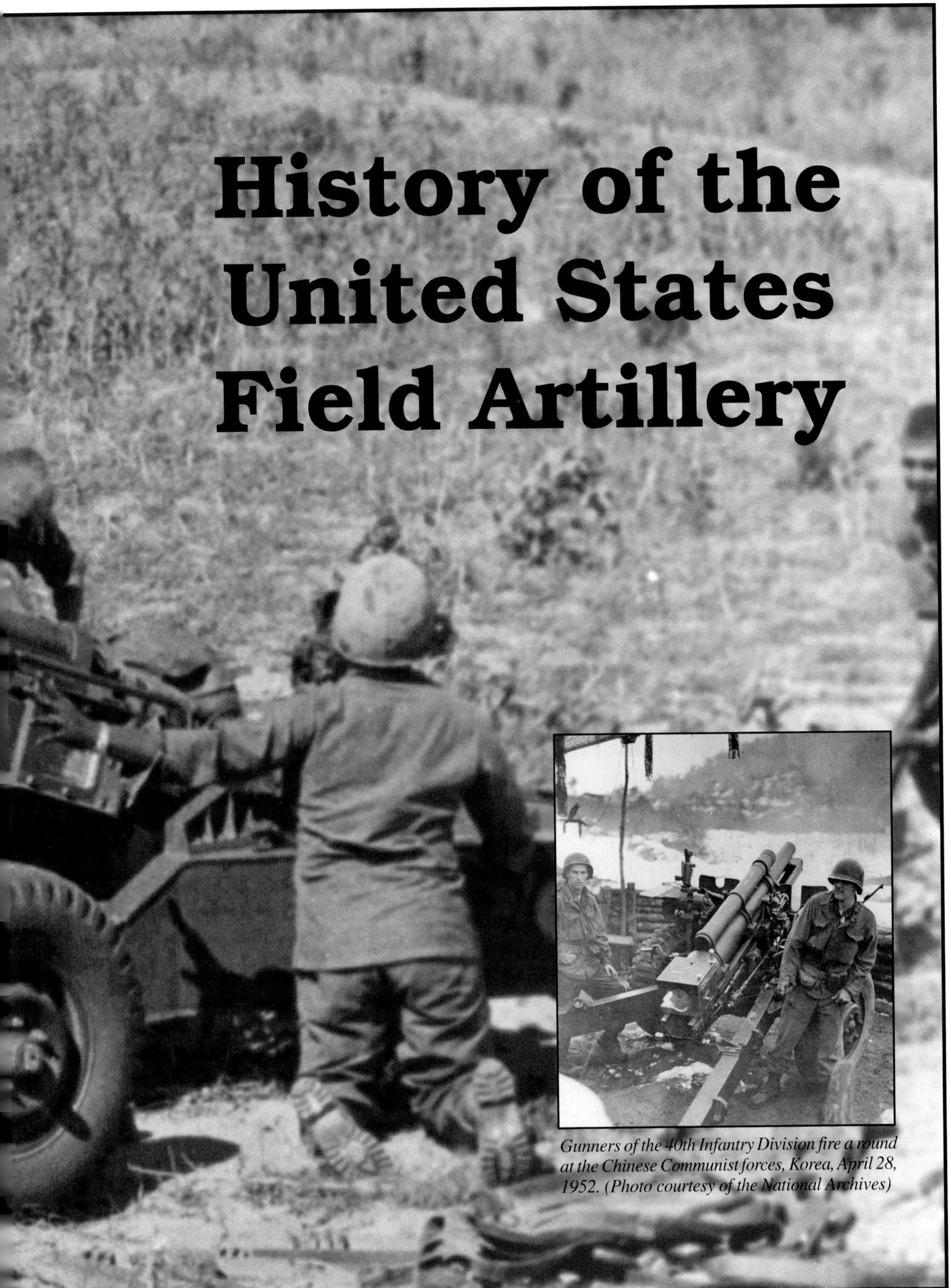

Gunners of the 40th Infantry Division fire a round at the Chinese Communist forces, Korea, April 28, 1952. (Photo courtesy of the National Archives)

An Overview of Artillery Evolution

This book is primarily a history of field artillery as employed by the United States armed forces. We shall see that this branch has answered the call to arms in every major war and several low-intensity conflicts, and has been significant in most and decisive in many.

From small, smoothbore, muzzle-loading cannon, to today's rifled, breech-loading tubes, to rockets and missiles, the 222-year story of American artillery is at once one of evolution of technology and equipment and of employing this equipment in battle. Artillery is a broad concept that refers not only to the weapons, but also to the men that transport, load, aim and fire the weapons. In the United States, artillery is usually a weapon not hand-held, that fires ammunition having a diameter larger than one inch (25.4-mm) and includes missile and rocket launchers.

Various kinds of siege machines, mostly rock-throwing catapults and the ballista (a large cross-bow type of missile thrower) are known from pre-Christian times, but the development of firearms, including large guns, depended on the invention of gunpowder. Gunpowder, originally a mixture of sulfur, charcoal and saltpeter (potassium nitrate) and called black powder, appears to have originated in China early in the Christian Era and was known to Europeans by 300 AD. By 1000 AD, the Chinese were employing explosive mixtures in bombs and grenades, and in the 1200s, black powder was introduced in Europe.

Very soon, machinery was invented to use the explosive mixtures to hurl projectiles at an enemy on the battlefield. There are references to the use of wooden cannon (bombards) in addition to wrought-iron and cast-bronze pieces in the middle 1200s, and 200 years later, the French and English were using small cannons against each other in the Hundred Years War. These small weapons were made of iron or bronze and fired iron or lead balls. The French and Italians soon began mounting light artillery pieces on wagons, achieving some degree of mobility. Mortars that could fire high-trajectory projectiles over castle walls soon followed.

During the 1500s and 1600s, several important developments in artillery and its uses occurred in Europe. In 1509, the Duke of Ferrara destroyed an entire Venetian fleet on the Po River in northern Italy by cannon fire. An Italian mathematician, Niccolo Tartaglia, founded the science of ballistics in 1537 by mathematically describing the trajectory of a cannonball. Exploding, hollow-cast shells came into use during the 16th century. By 1630, the Swedes had developed methods for employing artillery on the battlefield in support of the infantry. They increased the number of guns per unit of infantry to six guns per 1,000 men. King Gustavus Adolphus of Sweden organized the first regiment of artillerymen and established the first formal schools of artillery instruction as early as the 1620s. The Swedes also appear to have been the first to use cartridge ammunition, that is, bags of gunpowder rammed into the cannon instead of ladling in loose powder. By the end of the century, the Dutch had developed the medium-trajectory howitzer.

By the beginning of the 18th century, all the major countries in Europe realized the importance of artillery as both a siege and battlefield weapon and organized permanent artillery regiments and companies. At this time, artillery pieces were identified by the weight of the shot they propelled, so there now appeared in the literature references to 6- 9- and 12-pounders, for example. In England, modern ordnance theory and practice was developed by the 1740s by Benjamin Robins, and later, small cannons (carronades) were designed and built to arm ships of the Royal Navy. About this time, the word "gun" came into common use rather than "cannon" for pieces that fired a projectile in a relatively flat trajectory. Before the century closed, the Prussians had developed a "horse artillery," that is, artillerymen who rode with the cavalry, as opposed to slower "foot artillery," artillerymen who walked beside the horses hauling the heavy field pieces. Napoleon of France sent his artillery in front of his infantry to open large holes in enemy infantry lines with grape shot (clusters of small iron balls) and canister shot (a shell case filled with metal fragments). In the 1760s, Jean Baptiste de Gribeauval of France completely overhauled French artillery, redesigning virtually everything associated with artillery including organizational reforms, guns, wagons and ammunition. By the 1770s, French artillery became a model for the armies of Europe.

During the late 1800s many developments were initiated in the mechanics of artillery pieces, which we still see today: breech-loading mechanisms through which the projectile is loaded into the gun at the breech end rather than into the muzzle, which allowed an increase in the rate of firing; the rifled barrel, where the barrel had spiral grooves machined into its bore, which imparted a spin to the projectile and improved accuracy; recoil-absorbing systems that dampened the violent backward movement of the barrel (and carriage) on firing, and returned the barrel to its original position, thus allowing an increase in the rate of firing; smokeless powder that did not block visibility or identify the position of the gun or howitzer to the enemy; and advances in metallurgy such as the use of steel in gun manufacture.

Artillery weapons and artillery science advanced by giant steps in the 20th century, spurred by two great world wars and several smaller but equally violent conflicts and the constant threat of war in the latter half of the century. By World War I, the French had perfected a recoil-absorbing mechanism for their famous rapid-firing 75-mm gun. The Germans developed enormous weapons that could propel a projectile some 75 miles and a howitzer that could throw a 1,700-pound shell six miles. In the static trench warfare of World War I in France, artillery barrages of enormous firepower were common at the front and artillery tactics like the "walking barrage," in which a forward-moving rain of high explosives was laid down by artillery just ahead of the attacking infantry, were honed to perfection.

Such giant guns of World War I were of little use in World War II for the latter was a fast-moving war that required smaller, more mobile weapons. The 105-mm howitzer and the 155-mm gun and 155-mm howitzer, both towed and self-propelled, were the mainstay artillery pieces in America's war in Europe and the Pacific.

Today the lines separating artillery from other forms of armament are more blurred: rockets and missiles have joined the more conventional cannons and howitzers as artillery weapons, and many larger weapons in modern armies today are mounted on an armored, wheeled or tracked self-propelled chassis in addition to the towed pieces. Newer innovations include projectiles with submunitions that "home-in" on a target when dispensed in the target area. With such devices, pin-point accuracy over long ranges is possible.

With this brief evolutionary survey of artillery, we can now trace America's role in the development and use of artillery as a weapons system and as a specific branch of the armed forces. What follows will be a chronological history from near the beginning of the nation, using America's major wars as way points.

"America's First Field Artillery" by Joyce Kreafle

Artillery in Colonial America

The early American Colonists developed a militia to combat not only Indian attacks, but also threats by French and Spanish troops as Colonists, who were British subjects, expanded toward territory in the New World claimed by France and Spain. Joining the militia was obligatory, although some towns also organized a volunteer militia. But it soon was apparent that artillery was difficult to use against the hit-and-run tactics of the American Indian, and was very difficult to haul through dense forests and over rocky terrain. The musket was the weapon of choice in this guerrilla-type warfare. Cannons were not entirely ruled out, however. The British used artillery against static fortifications in siege operations along rivers and the Atlantic coast and to defend frontier outposts. All through the 1500s and 1600s, the British and French mounted cannons in their forts and other strongpoints from Florida to Canada. In 1638, only 18 years after the Pilgrims landed at Plymouth, the Ancient and Honorable Artillery Company of Boston, Massachusetts, was formed in that city to train the Colonists in fortress defense and bombardment.

But more than 100 years passed before the New England Colonists, under William Pepperrell of Maine, were ordered to a major engagement with the French (one of the battles of the four-year King George's War) when a French fort near Louisbourg on Cape Breton Island (now part of Nova Scotia) was besieged with smoothbore, muzzle-loading cannons and mortars for seven weeks. The French had been raiding New England shipping using this fort as a base of operations. With help from the Royal Navy, the French finally surrendered the fort to Pepperrell's band in June 1745. The Colonists had survived their first trial by fire.

The ensuing French and Indian Wars (1755–1763) pitted the British regulars and the American Colonists against the French and their Indian allies. Artillery played a very minor part in these wars with the exception of the British attack on Quebec in 1759, where two 6-pounder, canister-firing cannons tore holes in the French infantry lines. By the beginning of 1761, the French military forces had been effectively ejected from North America. But British problems in North America were just beginning; through the next ten years the Colonists became increasingly disenchanted with British laws that were often arbitrary and their enforcement arrogant. Among these were the Sugar Act, a duty on imported molasses, the Stamp Act, a tax on all legal and other printed documents, and the Quartering Act, requiring Colonists to temporarily house and provide for British troops. These taxes helped to meet the cost of maintaining British troops in the New World. Most of these were later rescinded by Parliament except for a tax on imported tea passed in 1773, and this one started the trouble. On December 16, 1773, in an unheard-of rebellious act, Bostonians, thinly disguised as Indians, dumped 342 chests of tea into Boston Harbor from British ships.

The American Revolution 1775–1783

In retaliation for the tea-dumping incident, the Intolerable Acts were passed by Britain which, among other indignities, closed the port of Boston to all trade. The other colonies rallied around Massachusetts, and a Continental Congress was formed and met in Philadelphia, in September of 1774, to express its displeasure with Britain. The British sent 3,000 troops to Boston under General Thomas Gage to collect arms from the Colonists and on April 19, 1775, a contingent of 700 regulars was attacked by armed Colonists at Lexington, Massachusetts, and were then routed at nearby Concord and forced to retreat to Boston. In this debacle, the British suffered 73 killed and 174 wounded and the Colonists 49 killed and 41 wounded. As many as 4,000 Colonists joined in this chase to Boston. Thus began the American Revolution, a war that Britain, with its massive resources in arms, men and supplies, should have put down in a few months, if not weeks, but it lasted six and one-half years. The Colonists began arming themselves and men from all the New England colonies began to converge on Boston.

Two months after the Concord confrontation, Massachusetts organized an artillery regiment under Colonel Richard Gridley, who had bombarded the French at Louisbourg 30 years before. Each of the nine companies in this regiment was made up of about 60 men, and each company had four guns. These few cannons were an assemblage of European-made weapons of a variety of calibers. Later in this year, Pennsylvania and Massachusetts foundries would start making guns, howitzers and mortars for the patriots, and the French would supply more artillery pieces. American artillerymen complained constantly of the shortage of powder.

The Colonists occupied Breed's Hill on Charleston Peninsula overlooking Boston on the night and early morning of June 16 and 17, 1775, although the orders were to occupy the adjacent, and higher, Bunker's Hill, and brought six field pieces to bear on the British. Newly arrived General William Howe, acting under the British commander at Boston (General Gage), sent his infantry and eight 6-pounders across the Charles River in boats from Boston to Charleston. Warships in Boston Harbor bombarded the Colonist's position on the hill, but the rag-tag Colonists kept building their earthwork position. On June 17, after a canister bombardment of the Colonist's hill position, Howe ordered his infantry to attack. The Colonists allowed the Redcoats to approach to within 15 paces (about 40 feet) and then opened a devastating fire with muskets. Howe's first attack was cut down, and the survivors ran to the beaches. Howe's second frontal attack in the late afternoon also failed. On the third attack, the British troops reached the hill fortifications and, now out of ammunition, the Colonists were forced to retreat. Howe had placed his field pieces on the flank in this third attack and enfiladed the Colonist's position with canister. With no means to fight back, the defenders had no choice but to give up the hill and five of their six artillery pieces with it. American losses, estimated at 500 casualties, were light but it was a disaster for the British. In the three attacks, 1,054 of the 2,200 infantry were killed or wounded, the heaviest British losses in any battle of this long war.

Colonial artillery was generally ineffective in this first major encounter with the British. In fact, two Colonial officers were charged with cowardice, and one was dismissed from the service. Colonel Gridley was nominally in command of artillery but he was not at Breed's Hill. Newly appointed on June 16, American commander-in-chief General George Washington removed Gridley for incompetence and the second Continental Congress appointed Henry Knox as Chief of Artillery of the Continental Army. Knox, 26, was well-grounded in artillery theory from his own reading and had received some training from British artillery instructors. The date of Knox's appointment was November 17, 1775, and this date is regarded as the beginning of the American Field Artillery.

Other Colonists under Ethan Allen and Benedict Arnold had attacked and captured Fort Ticonderoga in upstate New York in May of 1775, along with a large number of British artillery pieces. Twenty-one other field pieces were confiscated by the New York militia in a raid on a British battery on Manhattan Island in August 1775. Washington and Knox felt if they could amass enough artillery pieces they could again lay siege to Boston, and to that end, Washington ordered the artillery captured at Fort Ticonderoga be brought to Boston. Knox, himself, traveled to Fort Ticonderoga in December 1775 to bring the pieces to Boston. It was a Herculean task; 59 pieces — 13-inch mortars, howitzers and 9-, 12- and 18-pounder guns — were dragged on sleds and floated on barges for 300 miles from Fort Ticonderoga across the nearly roadless, snow-covered western Massachusetts to Boston. By the end of February 1776, the artillery pieces were in place on Dorchester Heights southwest of Boston and trained on the British garrison and ships in the harbor. Knox's bombardment began on March 5, and the buildings of Boston — and British morale — began to crumble under the withering fire. General Howe was defenseless against this assault. He spiked the vents of his own artillery pieces to prevent their use by the Colonists, and then vacated the city for Halifax, now Nova Scotia, on March 17. The next day

Washington entered Boston. It was a major victory for the Colonists, and it was achieved solely by Knox's artillery, not a musket ball had been fired. With Howe's departure, the British temporarily gave up all of New England to the rebels.

In the winter of 1775, Richard Montgomery and Benedict Arnold, with Washington's blessing, launched an ill-fated attack on British-held Canada. Montreal fell to Montgomery on November 13, and he joined forces with Arnold at Quebec and attacked that city on December 31, 1775. Lacking heavy artillery, the few 6- and 12-pounders under John Lamb were silenced by the British 32-pounders. The attack failed and British reinforcements forced an American retreat in the Spring of the new year, ending the Colonist's threat to Canada.

1776

With the Canadian debacle over, the British decided to send a flotilla of ten warships and 30 troop transports under General Henry Clinton and Admiral Peter Parker to the Carolinas as a show of force for the local supporters of the Crown. But this was to be a British debacle, for the commanders were unsure of their mission. In May, this flotilla of 40 ships headed south to Charleston, South Carolina, which they reached on June 4. The city's inhabitants, who had not yet involved themselves in any hostilities against the Crown, immediately started preparing their defenses, just in case this flotilla should attack them. On June 28, Clinton and Parker decided to do just that; it was a mistake from the beginning. The South Carolinians had fortified two islands at the harbor entrance, and these were Clinton's objectives. An infantry landing on one was a disaster as the British Redcoats were cut down. Then some of Parker's ships ran aground or blundered into backwater channels and out of the fight. Parker's cannon balls were absorbed by the fort's spongy, palmetto-log shields. The men on the two defending islands slowly but surely sent ball after ball into the British ships until some were reduced to floating wreckage. By sundown, Parker withdrew out of range, and in a few days, set what sail he could for New York. South Carolina was in the Revolution.

Washington had no sooner taken Boston when he turned south with five regiments, and Knox's artillery, to set up defenses on Manhattan Island, New York. He had about 20,000 men in his command and, for the next three months, fortified New York and nearby Long Island. Among the untrained militia were seasoned veterans of the Delaware and Maryland regiments, and all northern colonies were represented.

On the morning of June 29, more than 100 British ships appeared off New York. Howe had returned from Halifax and had brought with him his infantry and 40 artillery pieces. He would not be embarrassed by this ragged American army of farmers and shopkeepers again. A month later more ships arrived, this time with hired German troops. Then Clinton and Parker arrived from the Charleston debacle. Howe's infantry now numbered about 32,000 men. This was an enormous attack force and its job was made easier, for Washington made the mistake of dividing, and so diluting, his forces between Manhattan Island and Long Island. The attacking ships had free passage up the Hudson and East Rivers and the Colonial army was now about to be trapped. On August 22, 1776, Howe's infantry landed on Long Island and, five days later, virtually destroyed the American left flank and center. Howe then decimated the American right with his artillery and Washington's 9,500-man army was saved only by the stand of the Delaware and Maryland Continentals, all professional soldiers. Washington realized his force on Manhattan Island was about to be outflanked and ordered a general retreat northward to Harlem and White

Plains where he dug in. In September, Washington mauled a British-Hessian (German) unit at Harlem, but by October found it prudent to withdraw his battered forces from the area with British General Charles Cornwallis in hot pursuit. In this general withdrawal, the Colonists used their artillery as siege or garrison pieces from behind redoubts, in an attempt to slow the British advance, rather than as infantry support weapons, as the British used theirs. The rout on Long Island cost the Colonists 1,000 casualties, and they lost 146 artillery pieces and 2,800 muskets to the victorious British. But America's only army escaped, after inflicting only minor damage on the British. The British occupied New York and held it for the rest of the war.

It was obvious to Washington that his artillery had to be reorganized and trained in battlefield tactics. To this end the Continental Congress contracted with the French for the purchase of field pieces — guns, howitzers and mortars — and asked for French instructors to come to America to teach the Colonists how to use them. By the spring of 1777, these French artillerists were hard at work training the Colonists how to use their cannon in the field with the infantry against their common enemy.

Congress started enlarging the Continental Army late in 1776. In December, it promoted Henry Knox to Brigadier General of the Artillery and authorized Washington to raise 110 battalions of infantry and three regiments of artillery. Each artillery regiment would have field, siege and garrison artillery and have three field-grade staff officers and twelve companies. Each company would be composed of three officers and 54 enlisted men. Although the regiment's structure was a composite of different kinds of artillery companies (field, siege and garrison), tactics stressed employing light, mobile cannons as foot artillery for close support.

As Washington's army settled into winter quarters in Pennsylvania, Knox organized two regiments of artillery, added a third Virginia regiment and later absorbed a fourth Pennsylvania regiment in the summer of 1777. Artificers (mechanics, maintenance and repairmen) were formed into companies for each of these regiments to maintain the cannons, wagons, mobile forges and all other artillery-related equipment in the field. American artillery was coming of age.

While all this was going on, Washington, on the stormy Christmas night of 1776, ferried 18 field pieces across the icy Delaware River and distributed them to his infantry battalions. This was an unusual concentration of artillery for his band of 2,400 men. Hessian-occupied Trenton was the target and Washington attacked at first light on the 26th. As the Hessians partied, Knox deployed his guns to cover all exits from the town. As the Germans stumbled into the streets when alerted to the impending attack, Knox opened fire and shredded the Hessian infantry with canister. The Continentals then attacked and in wild street fighting overwhelmed the enemy garrison in less than an hour. There was no need for Knox to maneuver his guns in this brief battle, for the fighting area was so restricted, but this was the first time the patriots had used artillery as an infantry support weapon. The British immediately sent reinforcements under Cornwallis to Trenton, but Washington avoided battle and withdrew back across the Delaware.

1777

The rebels crossed back into New Jersey on New Year's Day, 1777, and headed for Princeton, about 11 miles into New Jersey from Trenton. Here, on January 2, Washington struck Cornwallis' rear and took Princeton. Although this bold attack forced the British to give up most of New Jersey, Washington's small army was exhausted and he went into winter quarters at Morristown, New Jersey. Washington built up his army through the spring and summer of

1777, but they were ill-trained and had poor equipment. Most dangerously, Washington's officers were inexperienced with only a rudimentary knowledge of tactics.

By late spring, British General John Burgoyne began marching south from Canada, heading for Albany with his force of regulars. With 15,000 men, General Howe sailed from New York to the head of Chesapeake Bay, intending to march the 60 miles overland to capture Philadelphia. Washington deployed his forces from Philadelphia and headed south to meet this British force. It was the last week in August, and the battle would take place near Brandywine Creek. The two armies finally met on September 11. Howe struck the American right, commanded there by John Sullivan of New Hampshire, as Washington's cannons were engaged in an artillery duel in the center with a British diversionary force. Washington sent Nathanael Greene's Virginia troops to aid Sullivan, but it was too late. The American right held at first, then was slowly rolled back, but maintained good order. Then the British attacked Washington's left, and that side was forced back also. Washington's artillery fired furiously and slowed the Redcoat attack to a crawl, taking a terrible toll, but could not stop the advance. Howe finally stopped some 15 miles from Philadelphia, where Washington had retreated. It had been a bad day for the Colonists. Howe entered Philadelphia on September 26, but again, Washington's army escaped.

Farther north above Albany, New Hampshire and Massachusetts militia, under Major General Horatio Gates, had meanwhile attacked Burgoyne's force of British and Germans coming down from Canada, beginning with an attack on an 800-man German unit on August 15, which wiped out the 800 almost to the man. After two months of incessant guerrilla attacks by the Americans, Burgoyne surrendered his battered, half-starved army of 5,000 men at Saratoga on October 17. Artillery played an important part in the final days at Saratoga. The Americans had placed their cannon on heights overlooking the dug-in British troops and a rain of solid shot and canister soon made their position totally untenable. But still, the Americans were reluctant to use their artillery in the field in support of infantry maneuvers.

Howe and Washington met again on October 4 near Germantown and, in a see-saw battle that saw the Colonials winning at one point, ended it with a British victory. Knox's artillery and his cannoneers were superb; they were learning their lessons well, and not a battery broke ranks or was overrun in the wild melee. By November, Howe controlled all approaches to Philadelphia and in December, Washington went into winter quarters at Valley Forge. 1777 had not been a good year for Washington's army. But during this long winter, a professional army took shape under the strict tutelage of the Prussian Major General Baron Friedrich von Steuben, and by the spring of 1778, Washington was commanding an army of disciplined soldiers.

1778

Knox continued to train his artillerymen through the winter, introducing a formal routine of training. A 6-pounder, for example, was usually serviced by 14 men; six were assigned to manhandle and position the gun with prolonges (ropes) while the others sponged the hot barrel and doused latent powder sparks, loaded and rammed the charge, and aimed and fired the piece. Their duties were:

1-6 Handled and positioned the piece.
7 Sponged the bore and (after No. 8) rammed in the charge.
8 Placed ammunition in the bore.

9 Thumbed the vent (a passage leading to the main propellant charge) at the breech end to prevent sparks from lodging in the passage while No. 7 sponged the tube and primed the vent with powder for the next shot.
10 Fired the cannon with a burning 'portfire' to the vent, the portfire being a paper cylinder of slow-burning powder or a burning quill.
11 Manipulated a 'handspike' (a bar used as a lever) to point the gun in the correct direction.
12 Held a bucket of water for No. 7's sponge, and a 'linstock,' a lighted wick to ignite the portfire (or to fire the cannon).
13 Handed off ammunition to No. 8.
14 Was in charge of the ammunition supply, handing ammunition to No. 13.

A fifteenth man was often present to hold the horses used to maneuver the cannon. Each man had to know and be able to perform the other's duties.

Also at this time, Knox reduced the confusing number of calibers in the artillery trains and determined the number and distribution of pieces:

To each brigade as battalion guns: 3-,4- and 6-pounders (four). General support for each brigade: 3- and 4-pounders (ten) and 6-pounders (ten), 5.5-inch howitzers (eight), 8-inch howitzers (four), 12-pounders (four), and 24-pounders (two). Reserve artillery: 3-, 4- and 6-pounders (30), 12-pounders (two), and 24-pounder (one). Siege artillery: 5.5-, 8-, 10- and 13-inch mortars, and 18-, 24- and 32-pounders.

In May 1778, Congress ordered the formation of four Continental Artillery Regiments totaling 44 companies. Colonels Charles Harrison (1st Regiment), John Lamb (2nd Regiment), John Crane (3rd Regiment) and Thomas Proctor (4th Regiment) were the Regimental Commanders.

Also in May 1778, France formally declared itself for the American side that in essence, declared war on England. The slow-moving Howe was replaced by General Clinton as British commander in North America. In June, Clinton was ordered to withdraw from Philadelphia for the French could be expected to strike the coast of America at any time. The last major battle in the north, army against army, Washington against Clinton, occurred on a searingly hot June 28 when Washington, with more than 13,000 men, attacked Clinton's withdrawing column at Monmouth in north-central New Jersey. Clinton's forces were 11,000 strong with a 12-mile-long baggage train of 1,500 wagons. Knox placed his artillery on Washington's flanks and so decimated Clinton's rear that, even at musket range, the British commanders were unable to attack in force. In the end it was a draw, the British and Germans lost many men in attack after attack, and as darkness fell, the patriots occupied the battleground, but Clinton's force escaped in good order. This battle was fought using conventional tactics of the times and the Colonists showed they could fight and defeat the British on the enemy's own terms. Of this battle, Washington wrote, "The American Artillerist proved himself a match for the batteries of Britain; the enemy had done them the justice to acknowledge that no artillery could be better served." And in a letter to Congress he said, "We forced the enemy from the field and encamped on their ground." When Clinton arrived in New Jersey and started to ferry his column across the harbor to Manhattan, a French fleet arrived off New York, but did not attack. Washington then camped at White Plains, New York, to watch for Clinton's next move.

This battle at Monmouth is also remembered for the courageous actions of one Mary Hays McCauly, renamed "Molly Pitcher" by the men in blue. Mary Hays survived the bitter winter encampment at Valley Forge with her husband William Hays, but on June 28, she placed herself in the thick of the battle, carrying pitchers of water to cool the bores of Knox's hot cannons and to quench the thirst of the cannoneers. Between helping wounded soldiers to the rear, "Molly" took the place of her wounded husband and swabbed and rammed his battery's cannon. Washington made her a noncommissioned officer for this bravery under fire and "Sergeant Molly" became a legend in the Continental Army.

1779–1780

As noted above, the battle at Monmouth was the last major engagement between these contending armies in the north. Washington was no longer strong enough in the north to battle Clinton in the field, and subsequent engagements would shift south and be fought by detachments of smaller armies. Clinton was ordered south to conduct operations against the rebels. The use of artillery changed dramatically in these southern confrontations. Artillery was used sparingly for there was little of it. With the stalemate in the northern colonies, Clinton shifted his attacks to the south (He had captured Savannah in December 1778.). British forces under Clinton captured Charleston, South Carolina, in May 1780, after local artillery units forced him into a long, six-week siege of the city. Clinton then returned to New York leaving Cornwallis in command. Cornwallis formed a chain of armed posts in South Carolina and immediately the local militia under Horatio Gates began attacking them. The Americans were defeated at Camden, South Carolina, in August 1780, in a furious infantry and artillery melee where Gates lost his seven field pieces to the victorious British. A Tory force was destroyed at King's Mountain in October, although no artillery was involved. The Tory leader, Major Patrick Ferguson, had invented and patented a breech-loading rifle less than four years earlier; Ferguson was killed at King's Mountain. The rebels in the southern colonies used hit-and-run guerrilla tactics, a type of fighting they knew well. Among their leaders was Francis Marion, the "Swamp Fox," who proceeded to methodically destroy Cornwallis' outposts throughout South Carolina, one by one. In December 1780, Nathanael Greene was given command of the southern operations, replacing Gates. Greene had only 1,600 troops in this command, mostly untrained militia, with sparse elements of Benjamin Harrison's and Thomas Proctor's artillery.

1781

Battles were won and lost, beginning in 1781, by both sides, but the rebels could live off the land and the British became sorely depleted in men and supplies even though they won many engagements. A British dragoon unit was soundly routed at Cowpens, South Carolina, by Dan Morgan's patriot militia in January 1781, but one British victory was at Guilford Courthouse, North Carolina, in March 1781, which began with an artillery duel that decimated the infantry of both sides. Greene eventually had to withdraw before Redcoat bayonet charges, but so did Cornwallis, for he had only shreds of an army left. Cornwallis, who saw Greene "...as dangerous as Washington," was replaced in April by Lord Francis Rawdon as overall Commander. Greene continued to attack into South Carolina at outpost after outpost as 1781 wore on, and on September 8, the last battle was fought in South Carolina at Eutaw Springs, 45 miles northwest of Charleston. Greene had two 3-pounders and two 6-pounders, which he used to force the British into a brick building. But the artillery pieces were not heavy enough to breech the brick walls, and the Redcoats soon launched a bayonet attack and drove Greene's troops from the field. The British lost more than 800 men, including prisoners, and were forced to retire to Charleston with their wounded.

These long months of attrition had worn down the British and they would fight no more in the colonies south of Virginia.

Meanwhile, on being relieved by Rawdon, Cornwallis had moved his forces from the Carolinas to Virginia, first to Portsmouth and then to Yorktown. The traitor, Benedict Arnold, was now in the employ of the British as a Brigadier General, and was assigned to subdue the rebel militia in Virginia, which he readily did. Earlier in 1779, a British amphibious force from New York had already sacked many coastal Virginia towns, destroying much of the rebel militia's supplies. There was little opposition to these British raids of destruction for Governor Thomas Jefferson had already sent most of Virginia's militia south to fight with Greene. But now the showdown was building.

Washington, at Tappan, New York, learned of Arnold's presence in Virginia and a plan was hatched in February 1781 to send the handsome Frenchman, the Marquis de Lafayette, with the rank of Major General, south with 1,200 New England and New Jersey troops, disciplined and battle-tested. It was also agreed that the French would sail from Newport, Rhode Island, with an additional 1,200 men. This small band was to defend Virginia and, if they could, capture the traitor Arnold in the process. The British, anticipating a battle, sent 2,600 men to Virginia under William Phillips to counter Lafayette. Cornwallis brought another 1,500 men with him from the Carolinas in May, and Clinton sent an additional 1,500 infantry from New York. Milling around Portsmouth and Yorktown, now, were more than 7,000 Redcoats, their leaders spoiled for a final fight with the rebels. Cornwallis was then given command of all these forces. Lafayette ignored these formidable numbers and continued his march south, finally reaching Richmond in May. His band had swelled to 3,000. But Lafayette was no fool; he knew he could not win a major battle with the British, so chose small skirmishes to bleed small British units, wherever he could find them.

When Cornwallis sortied into the Virginia countryside, destroying any rebel stores he could find and searching for Lafayette, the Marquis simply fell back, avoiding all contact with Cornwallis, which would have been disastrous for his small band of rebels. In June, Anthony Wayne arrived in Fredricksburg, Virginia, with three regiments of Pennsylvanians from Washington's army, and finally, as other units filtered in, Lafayette felt strong enough to carry the war to the British. His little army had swelled to some 4,500 militia and Continentals. Minor skirmishes erupted near Williamsburg, but Cornwallis was unperturbed; he outnumbered Lafayette and Wayne, and the Royal Navy could bring him reinforcements at any time. Early in July, Cornwallis set a trap for the rebels, luring them to attack a small, exposed force of Redcoats. Wayne came on with 900 men and then the wily British General released 3,000 hidden troops to the battle. Wayne was lucky to escape with most of his command intact.

Cornwallis then received a series of conflicting orders from Clinton in New York, first, wanting 3,000 men sent to him on Manhattan, then to Philadelphia, then it was New York again, and finally, on July 20, 1781, these men were to stay in Portsmouth, Yorktown or Old Point Comfort. Exasperated, Cornwallis chose Yorktown, sailing his entire force up the York River from Portsmouth. He attacked and seized Yorktown and also Gloucester Point on the north side of the York River. In August, Washington received intelligence of the movement of the French fleet from the West Indies to the Virginia coast and started a march south 450 miles to Virginia. He had 7,000 men in his command, 5,000 of them French regulars. The move was in secret, and Clinton, in New York, did not learn of it until early September. A few days later French Admiral Francois Comte deGrasse put ashore 3,000 French troops on Virginia's Jamestown Island, joining Lafayette and Wayne. Clinton sent southward some British ships to engage the French fleet, but deGrasse's force of 24 ships and some 1,700 cannons summarily turned away the British force. Cornwallis was now surrounded by land and sea. It was at this time, September 8, that Greene surprised a 2,000-man British force at Eutaw Springs. Washington established his headquarters at nearby Williamsburg and began the siege on Cornwallis at Yorktown on October 9, 1781. The outcome of the battle was never in doubt. Cornwallis commanded about 6,000 men who had no way to escape their encirclement. Washington commanded 8,800 militia and Continentals and nearly as many French troops. Washington, himself, fired the first cannon shot at Cornwallis' dug-in troops and then the field and siege artillery of the 2nd and 4th Continental Artillery Regiments, together with French artillery, took a terrible toll on the British. British cannon responded, but it would not be enough. On October 14, Colonel Alexander Hamilton led an attack on a British strongpoint and conquered it. On October 17, Cornwallis called for talks and formally surrendered Yorktown two days later. Negotiations with the British dragged on for nearly two years until the formal treaty was signed at Paris on September 3, 1783, ending the Revolution.

The Revolutionary War had seen the birth and subsequent transformation of American artillery into a well-trained fighting force. But at this time, the Europeans were using field artillery as weapons exclusively in support of infantry maneuvers. The patriots in America used their artillery weapons for all purposes: field, garrison and siege. They did not recognize a specific branch of field artillery as did the Europeans at this time.

Artillery After the Revolution 1783–1812

For 31 years following the final battle of the Revolution at Yorktown, the United States was at relative peace with the European powers, if not so at home. Raiding parties of various Indian nations constantly attacked the westward-pushing new United States citizens. Congress tried to control this frontier by raising eight companies of infantry and two of artillery from the northern states in 1785. The Indian attacks could not be controlled, so Congress enlisted more men in 1786. But Congress could not find funds to support even this small army of 2,000 men and discharged all but eight infantry companies (about 600 men) and four companies of artillery (about 290 men). In 1787, Congress was forced to enlist more men to protect the frontiers of the new nation. The Indian raids continued into the 1790s, but the hit-and-run tactics of the forest Indians seemed unstoppable.

Artillery was of almost no use in these woodland skirmishes. Small pieces were used on occasion to bombard Indian villages and were used as garrison weapons in a string of forts constructed westward to the future state of Indiana. The thick forests denied artillery clear fields of fire, and so the American commanders often did not transport artillery pieces with the infantry. Further, there were few qualified artillerymen since the end of the war, for they had been used as infantry and over the years had lost their expertise. The canister-firing, 3-inch howitzer saw some success against groups of Indians, but even transporting this light weapon (about 220 pounds) proved a chore in the rocky and wooded Ohio terrain. This state of affairs continued into the new century, but eventually, the Indian attacks decreased in intensity.

Beginning in the 1790s, trouble with Great Britain started again as British ships raided the Atlantic coast. The new nation responded by manufacturing 12-, 24- and 32-pounder iron cannons and placing them behind earthwork redoubts as coastal artillery. Congress created the Corps of Artillery and Engineers in 1794, and all existing artillery regiments were absorbed into this new Corps. An artillery school was also established at West Point, New York, but almost no funding made it a school in name only. By the end of the 1790s, the United States could boast two regiments of 28 artillery companies for both coastal and frontier duty. But in training and equipment, these companies were no match for artillery of the European countries in the early 1800s, especially those of France's Napoleon Bonaparte. Horse artillery was adopted in Europe in the middle 1790s, and it eventually became the prime, heavy-hitting arm of the infantry on the battlefield. Napoleon improved the ability to mass fires by removing control of artillery from battalion commanders and concentrating light weapons at the division level and heavy weapons at the corps level. Ranges of field guns at this time rarely exceeded 1,000 yards. Fixed ammunition, with its shot and propellant in a single unit, was available at this time, but seldom used except when rapid fire was necessary. Semi-fixed ammunition, with the projectile and powder bag loaded separately, was still the rule.

Other significant changes occurred in the 1790s. Britain's Captain William Congreve redesigned the trail carriage, limber and caisson, resulting in a "mounted" artillery. Congreve replaced the double, wooden-beam trails of the cannon carriage with a single-beam trail, giving the piece a shorter turning radius. Second, he designed a limber, the wagon to which the gun is attached for transport, with a readily accessible ammunition chest that could also seat part of the gun crew. Third, he redesigned the caisson or ammunition wagon to seat the rest of the gun crew. Another development in the early 1800s was the adoption of rocket artillery by the British. Although accuracy was always a problem, the rockets had a range of up to 3,000 yards and could be tipped with solid shot and canister. Other European armies quickly adopted the new weapon.

All these European developments stimulated the United States to upgrade its own artillery so it would not fall behind potential enemies in Europe. But all too familiar in the new nation, Congress had no money for improvements in artillery or anything military. President Thomas Jefferson, in 1802, created the separate Corps of Engineers and, in spite of the protests of a series of strong War Secretaries, simultaneously gutted the infantry, artillery and cavalry as the possibility of a war with European nations became more remote. A debate over the virtues of iron versus bronze cannons slowed artillery manufacture in the early years of the 1800s and a shortage developed, but improvements in carriages, limbers and caissons made the heavy pieces more maneuverable. There remained a serious lack of trained artillerymen since the frontiers of the nation required primarily a sharpshooting infantry.

War with Britain
1812–1815

This war, the War of 1812, was America's strangest, even most tragic war. Because of the slowness of sailing ships, the only means of transatlantic communication, it need not have happened. England and France's Napoleon were at war, and both attempted to blockade the other's ports. In doing so, each interfered with the shipping of neutral nations, the United States being one. Britain ordered all ships bound for France to stop at a British port for inspection and payment of fees. Napoleon then notified all noncombatants that if they complied with this fiat, he would seize such ships. Britain then began stopping American ships on the high seas to search for British seamen who had deserted from the Royal Navy. Incidents occurred, one in 1807, in which American ships were fired on and American seamen were impressed into British service "by mistake." Through 1811, many attempts were made to reach some agreement with the British to change their policies against neutral nations. Little progress was made. On June 18, 1812, Congress declared war on Great Britain. Tragically, two days earlier, Parliament had repealed all the offending policies that were the reason for the declaration of war. The news of the repeal did not cross the Atlantic until well after hostilities had begun on the American continent. The young nation had embarked on its first formally declared war with the strongest power on earth — England.

At the start of the war, the United States Army was composed of seven infantry and two artillery regiments and one dragoon regiment. These units were scattered up and down the western frontier and at coastal defense forts. Two more artillery regiments were authorized in June when war was declared. The preceding February, a light artillery regiment was formed with two batteries (eight guns) and sent to Niagara, New York. Later in the fall, three additional light batteries were added and sent to Plattsburgh, New York.

1812

The Americans immediately mounted an attack on Canada from three points: Detroit, now Michigan, the Niagara River and the Lake Champlain area. Most of Britain's six regiments of infantry and eight artillery companies, were spread out along the eastern Great Lakes. All American attempts met with disaster. General William Hull's 2,000 men crossed the Detroit River into Canada in August and were driven back to Detroit where the entire force was captured. General Stephen Van Rensselaer's New York militia of about 6,000 men crossed into Canada at Niagara on October 13, and some 400 of these attacked and occupied Queenston Heights, Ontario, about seven miles below the Falls. American artillery of two light field pieces under Lieutenant Colonel Winfield Scott and two 18-pounders under Captain John Lovett were ferried across the river and used effectively against British infantry and artillery, and the Americans captured an enemy 18-pounder. Van Rensselaer also had 6-pounders, but no boats to ferry them. A British counterattack drove them back to the American side of the Niagara River, capturing 900 Americans. It was similar at the Lake Champlain attack. The militia reached the Canadian border, but then refused to cross into Canada. The little army returned to Plattsburgh.

Ontario, Canada, north of Detroit, in the Battle of Moraviantown, regaining the Northwest Territory. The British were supported by about 600 Indians under Tecumseh who was killed in the battle. Also in the fall, Montreal was attacked by the Americans, but this attack failed. At year's end, the British captured Fort Niagara and burned Buffalo, both on the American side.

1813

In January, a detachment of Kentucky men was defeated by a British unit and Indians at Frenchtown, now Monroe, Michigan, on the Raisin River, south of Detroit. The Indians massacred the wounded Americans after the battle. In April, York (Toronto) was captured by the Americans and held for a short time. Six-pounders and 5.5-inch howitzers were used in this attack, essentially as siege weapons. President James Monroe assigned General William Harrison the task of regaining the Northwest Territory that Hull had lost at Detroit. Harrison spent the winter at Fort Meigs, south of Detroit, and fended off British attacks there in May 1813, with his 6-, 12- and 18-pounder garrison weapons.

On Lake Erie, at its far western end, Commodore Oliver Perry defeated a British squadron of six ships on September 10, in one of two American naval victories. Perry had built two brigantines on the lake, each armed with 20 guns; in addition, he had captured a British brigantine and had six small schooners with one or two guns each. A sharp surface battle ensued and the outgunned British finally surrendered their fleet to Perry. Perry wrote to President Harrison,"We have met the enemy and they are ours."

Finally, feeling strong enough after Perry's naval victory on Lake Erie in September, Harrison, with 3,000 men, attacked and defeated the retreating British on the Thames River in

1814

England's war with France ended this year, and Britain could now turn her attention to the irritation on the American Continent. England immediately sent some 18,000 seasoned troops to Canada and so ended all American hopes of any victory there. In the spring of 1814, Colonel Winfield Scott began intense training of infantry, artillery and cavalry near Buffalo, New York. Feeling confident in his small army of two brigades, Scott, under the command of Major General Jacob Brown, crossed the Niagara River in July in search of the British forces. On July 5, the British attacked Scott's brigade at Chippewa and his 12-pounders, commanded by Captain Nathan Towson, returned canister to the attacking British infantry. Two other 12-pounders picked up the action. The enemy charge was stopped in its tracks and Brown's infantry counterattacked and routed the British troops. This was an open battlefield engagement and the artillery was used exclusively in direct fire against enemy infantry. Scott had trained his artillerymen superbly.

About three weeks later, Scott with 1,300 men, found the British in the field again, about 2,800 strong, at a place called Lundy's Lane on the Canadian side of the border and about one mile from Niagara Falls. An artillery duel, British 24-pounders against Towson's 12-pounders, began the fight, but the battle soon became a man-to-man melee, and artillery then could not be used by either side to support its infantry.

General Brown marched in his 2,700 troops from Chippewa, only a few miles away, and the battle continued into the night. American artillery fire early in the battle was answered by British rocket barrages in addition to their field pieces. The British and the Americans mauled each other badly, and both eventually withdrew from the field. The battle at Lundy's Lane was a draw, but it ended all incursions into Canada by Brown's troops. The foot (field) artillery earlier in the battle had been used effectively by the Americans, and these batteries served strictly as field support to the infantry for the remainder of the war.

The following September, 11,000 Redcoats descended from Canada and fought their way to Plattsburgh, but there they were stopped. On Lake Champlain, Thomas Macdonough defeated a British fleet on September 14, 1814. The British in this battle had four armed ships and some dozen rowed galleys to protect the invading British army coming south from Canada. Macdonough commanded four armed ships in addition to ten galleys. On the lake, Macdonough forced the British to approach head-on, then turned his ships broadside and raked the oncoming boats (akin to the famous naval maneuver called "crossing the T"). The small British fleet surrendered and the British land army withdrew to Canada. By Perry's victory on Lake Erie a year before, and now Macdonough's victory, American control of the western Great Lakes was assured, and enemy passage south from Canada via Lake Champlain and the Hudson River was effectively blocked.

The preceding month, the British under General Robert Ross, with cannons and rockets, had greater success in a fleet-escorted incursion into Chesapeake Bay. The artillery of General William Winder slowed the enemy advance at Bladensburg, Maryland, on the banks of a Potomac River branch with twenty 12- and 18-pounders, but the British advance could not be held. Enemy artillery and rocket barrages eventually forced the Americans to retreat to Washington. The British army marched into Washington with little opposition and burned many buildings in the town. Flushed with victory, Ross then took his army north to take Baltimore, but his luck ran out. Despite a great barrage of shot, canister and rockets, he could not take Fort McHenry that defended Baltimore Harbor, and victory was not his. Ross withdrew from Chesapeake Bay. Francis Scott Key wrote the words to the *The Star Spangled Banner* during the bombardment of Fort McHenry. One more battle, far removed from those fought up to this time, was now to take place at New Orleans, Louisiana, and would be the last major engagement of the war. Unknown to the combatants, peace terms had been agreed to some 15 days earlier in Ghent, Belgium. It would be a needless battle.

Major General Andrew Jackson commanded the American forces at New Orleans and had plenty of warning of the impending British attack. Jackson fortified all approaches to the city and rendered many impassable. A brief clash between the two armies occurred on December 23, 1814, which prompted Jackson to build a line of defense on the east bank of the Mississippi River. He positioned his artillery of calibers ranging from 6-pounders to 32-pounders and pointed them in the direction he guessed the British would select for their attack. Another, but smaller, battery was placed on the opposite bank to protect his flank.

1815

The British opened the battle on New Year's Day, 1815, but Jackson massed his fire on both enemy artillery and infantry, and forced the British to withdraw out of range. After reorganizing their forces, the British attacked again on January 8, this time with more determination. Jackson's 12- and 18-pounders shattered both flanks of the attacking British infantry on the east bank, but the enemy attack on the smaller west bank battery was more successful. A superior enemy force overran the American position, but British losses on both sides of the river were too great to take advantage of the victory. The enemy withdrew from the field and the battered British forces embarked on the ships that had brought them. As noted above, peace terms had been signed by the belligerents on December 24, 1814, nearly two weeks before the battle of New Orleans, and this battle, although a glorious victory for Jackson and his men, did not figure in the final outcome.

Artillery After the War with Britain 1815–1845

As seems always the case following a war, Congress immediately reduced the size of the Army in 1815. The number of infantry regiments was cut from 17 to eight, and artillery from four wartime regiments to one Light Artillery Regiment plus eight battalions of Corps of Artillery for coastal defense. Total size of the United States Army was set at 10,000 men. By 1821, this number had been reduced to 6,000 men, and all artillery was merged with the Ordnance Department. The number of artillery calibers was reduced following the War of 1812. The carriage and limber designs were retained, although these designs (the Gribeauval system) were now obsolete in Europe. Iron in manufacture of the artillery pieces was retained in America.

With no obvious enemy to America from overseas, and money in short supply, little was done by the Congress to catch up to modern artillery developments in Europe. Constant bickering continued between Congress and the War Department on the best artillery system to use. All through the 1820s and 1830s, guns, carriages, limbers and caissons were tested, mostly half-heartedly, without arriving at a system for artillery. Iron guns exploded in tests and the Gribeauval carriages and limbers were slow and cumbersome to maneuver, but nothing was done. Artillerymen were still being used as foot soldiers on the frontier, and there were no viable schools to train artillerymen. Sometime during the 1820s, Army artillerymen were given, or took upon themselves, the nickname "Redlegs" in reference to the two-inch red stripe on the trousers of artillery uniforms of the day.

Limited funds and a succession of Indian wars, in which artillery was sparingly used, militated against developing a healthy artillery branch. This state of affairs was about to change. Disputes arose with England over the boundaries of the Oregon territory and with Mexico over the boundaries and status of Texas. The Indians were now challenging the Army on the open plains against whom artillery might possibly be used effectively. In 1839, the War

Department undertook a sweeping study of American artillery with the intent of bringing it up to date to match that of the European powers.

In 1841, the Secretary of War ordered all new field pieces be of bronze, and a modified French system of carriages, limbers and caissons was to be adopted. Further, guns were to be 6- and 12-pounders and howitzers of 12-, 24- and 32-pounder caliber. In addition, a 12-pounder mountain (pack) howitzer was approved. Fixed ammunition became popular by the end of the decade, and time-fused ammunition, invented in Belgium, was adopted. Friction primers with lanyards began to replace portfires and linstocks in the 1840s, and finally, the Army organized and equipped field batteries of artillery as a tentative step in separating this branch from coastal and siege artillery. A school was established at Fort Washington, Trenton, New Jersey, to train new artillerists. Gun crew duties were reassigned resulting in the reduction of the crew size from 14 men to eight. The new bronze artillery pieces represented the latest advances in smoothbore technology, but in Europe, the more accurate rifled pieces were threatening to make this technology obsolete.

The United States Army at this time, still dispersed its artillery widely on the western frontier and still employed artillerists as infantry. These practices discouraged developing effective gun crews, and the policy of rotating artillerists among siege, garrison and field duty persisted and militated against development of the expertness that comes with specialization. Still, there were a few well-trained and effective gun crews held together by a handful of forward-looking officers. All batteries, at this time, were attached to either a brigade or a division whose commanders controlled their fire. This was the state of American artillery by the middle 1840s as disagreements heated up with Mexico over the status of Texas.

The Mexican-American War 1846–1848

The Texas territory had declared its independence from Mexico in 1836, although Mexico refused to recognize this declaration. President James Polk regarded the southern boundary of Texas at the Rio Grande River, while Mexico considered the boundary at the Nueces River, which at one point is more than 150 miles north of the Rio Grande. In spite of warnings from Mexico, Texas was admitted as the 28th state on December 29, 1845, following an offer by the United States to buy from Mexico large tracts of land in the west and southwest. A revolution was in progress in Mexico at this time, and neither of the contesting governments there would see Polk's emissary.

1846

Fearing an invasion, Polk ordered General Zachary Taylor at Corpus Christi to take a 3,000-man force to the Rio Grande. Taylor no sooner reached the river in April 1846, than a Mexican force crossed into Texas and skirmished with Taylor's band. On May 8, Taylor was scouting territory between his camp and the Gulf of Mexico when he ran across a large body of the Mexican army on an open plain near Palo Alto, Texas. Taylor was outnumbered three to one. In Taylor's force were three batteries of artillery, two with four 6-pounders (Major Samuel Ringgold's and Lieutenant James Duncan's) and one with two 18-pounder guns (Lieutenant William Churchill's). Taylor ordered both 6-pounder batteries out in front of his infantry, and these batteries, along with the 18-pounders, decimated the Mexican center with canister. The Mexican General, Mariano Arista, then decided on a flanking cavalry charge but Ringgold's and Duncan's artillerymen stopped these attacks as they were getting underway. As the infantry closed to within range, American small-arms fire wreaked havoc on Arista's army as American artillery continued to tear huge holes in the Mexican infantry, cavalry and artillery. Arista was forced from the field; he had been defeated almost by artillery alone. Ringgold was badly wounded in one of the artillery exchanges and would later die. The following day, the 9th, Taylor caught up with Arista again at Resaca de la Palma and Arista again felt the sting of American artillery. In this fight, infantry played a larger part, and forced the Mexicans back south across the Rio Grande and into Mexico. It was a red-letter day for the American artillery, for they showed artillerymen could not only maneuver their pieces effectively on the battlefield, supporting the infantry, but could load, aim and fire them rapidly and accurately. On May 13, the U.S. Congress declared war on Mexico.

Near the end of August, Taylor set out for Monterrey, 90 miles into Mexico from the Rio Grande, with orders from President Polk to occupy Mexico. With 6,000 men and added artillery, which he attached to his infantry brigades, Taylor attacked heavily defended Monterrey on September 24. He forced his way into the town and his men became engaged in a wild melee of street fighting. They could not sustain their momentum and were forced to withdraw. A second attack was equally bloody, but this time the Americans singled out and silenced Mexican artillery batteries and, in subsequent house-to-house fighting, forced the defenders to surrender. American artillery was not effective at Monterrey, for Taylor's 6- and 12-pounder pieces were too light to do any damage to the heavy stone walls of the city.

1847

Taylor was then required to send part of his force by sea, south to Veracruz, to aid in General Winfield Scott's drive to Mexico City. Taylor then marched 5,000 men and four artillery batteries to just beyond Saltillos, about 50 miles southwest of Monterrey, to a place called Buena Vista. Santa Anna, now President of Mexico, led an army of between 16,000 and 20,000 men to meet Taylor with the intention of forcing the Americans back across the Rio Grande. The battle started on February 22, with Santa Anna attacking the American left with light infantry. An American battery of a 6-pounder, a 12-pounder howitzer and a small captured 4-pounder Mexican piece, so decimated this attack that the Mexicans were forced to call off the charge for the day. A second attack on the morning of the 23rd was more successful. In spite of deadly shot, canister and shrapnel fired at 50-yard range, the battered Mexican troops continued advancing. Taylor brought up more artillery that fired at point-blank range, finally stopping the charge. Santa Anna, with little concern for the decimation of his right flank, now attacked the American center. Taylor shifted artillery to the center and again stopped the waves of enemy infantry. One more half-hearted charge by Santa Anna was repelled by massed artillery fire, and the Mexican

General was forced from the field. American artillery was superb in Taylor's campaign into Mexico. In the hands of trained crews, they were mobile, rapid-firing and accurate. With artillery and his numerically inferior forces, Taylor had secured northern Mexico.

On March 9, 1847, General Scott landed a force of about 10,000 men near Veracruz, far south along the Mexican Gulf coast. He layed siege to the city with shot, shell and rockets fired from surrounding hills. His artillery was light and not suited to siege, but served well to cover his attacking infantry. By the end of the month, Scott was in Veracruz. On April 8, Scott started his 200-mile overland march to Mexico City. Santa Anna's first attempt to stop Scott was on April 17 at a hilltop called Cerro Gordo on which Santa Anna had mounted four 4-pounders. Scott's engineers then led an artillery battery up a narrow path to a hill that overlooked the Mexican position. The battery consisted of three 24-pounders and rocket launchers. The rockets were a mixture of Hale and Congreve types. William Hale was an American inventor whose rockets were designed to spin in flight, thus improving accuracy. Sir William Congreve was British and invented, in 1806, a rocket with a range of 3,000 yards. Scott's attack was a textbook operation. The 24-pounders started their barrage on the enemy position on the morning of the 18th and then the infantry charged up the hill. Completely surprised, the Mexicans fled, pursued by Scott's infantry with the captured Mexican 4-pounders pouring shot into their backs.

Scott continued his attack west and northwest, taking Jalapa, about 60 miles from Veracruz, then Pueblo, only 70 miles from Mexico City, without a fight. On August 18, Scott's column, in attempting to bypass the heavily defended San Antonio, soon uncovered another heavily entrenched Mexican position at nearby Contreras, a few miles southwest of Mexico City. With supporting 12-pounders, howitzers and rockets, the Americans charged the position but were stopped by enemy artillery fire 400 yards from their objective. A Mexican cavalry charge was broken up, and that ended the fighting for the day. On the 19th, Scott's infantry charged again, stormed the strongpoint and drove the defenders out, capturing all 22 of the enemy's heavy artillery pieces. Another of Scott's columns attacked a Mexican force at Churubusco on August 20. American artillery supported the infantry attack but could not be maneuvered through the broken terrain. Once the American infantry had captured and silenced the enemy artillery, American 6-pounders were hastily brought forward to assault the enemy. Shot and canister rained down on the enemy until he finally surrendered.

Farther north, and just west of Mexico City, was a group of stone buildings called Molino del Rey, and on September 8, Scott opened his attack with 24-pounder shot that battered down the stone walls of the town leaving gaping holes for his infantry to pour through. The infantry took the town's artillery battery, fended off a cavalry charge, and American artillery shot and canister forced the defenders to flee east in the direction of Mexico City to a fortified castle called Chapultepec.

By September 12, four batteries of artillery had been positioned to almost completely surround the castle. Eight-inch howitzers, 6-, 12-, 16- and 24-pounder guns and a 10-inch mortar ringed the stronghold. The deafening barrage lasted the entire day until sunset, and on the 13th, the infantry attacked from the south and west in separate charges under cover of canister and shell fire. The fighting, at first, was hard and furious as Scott's men got closer to the castle. Then the defender's nerve broke and they fled eastward to the relative safety of Mexico City. Scott ordered immediate pursuit and, in fact, the artillery batteries were out front leading the charge into Mexico City. The Mexicans could no longer fight off this juggernaut of Scott's infantry and artillery, and they surrendered the city on the 14th. Santa Anna resigned after Mexico City fell, and a peace treaty with Mexico was signed on February 2, 1848. Again, the American artillery had proved its worth with Scott as it had earlier with Taylor in the north. As Dr. Boyd Dastrup, U.S. Field Artillery Center Command Historian, has pointed out, 6-pounder guns and 12-pounder howitzers were the heart of the Army's field artillery at this time and were organized into controllable batteries, but the heavier pieces — the 12-pounder guns and the 24- and 32-pounder howitzers — were not so grouped. Thus, this organization was tilted to favor mobility rather than massed firepower. Weapons used by the Americans in this war were 8- and 10-inch mortars, 12- 24- and 32-pounder howitzers and 6- and 12-pounder guns.

Artillery After the Mexican-American War 1848–1861

The thirteen years between the Mexican-American War and the beginning of the American Civil War was a period of rapid development in technology of artillery weaponry. These developments were centered around experiments with breech-loading pieces, rifled bores, use of steel in place of iron and bronze and cartridge ammunition. Europe led the way while America mostly watched, contributing little in this period, content that it had the best in smoothbore, iron and bronze, muzzle-loading artillery weapons, all of which would soon become obsolete. During this period, the U.S. Army reorganized its field batteries based on lessons learned in the war with Mexico. Beginning in 1857, the 12-pounder Napoleon gun/howitzer became the standard, with each siege battery made up of four 12-pounders plus two 24- or 36-pounder howitzers, and the infantry support battery of four 6-pounder guns plus two 12-pounder howitzers. Thus the problem of firepower versus mobility that plagued Scott's artillery in battles around Mexico City, was at least partially solved. But technology and not reorganization was the essence of artillery evolution the middle years of the 19th century.

Smoothbore cannon had effective ranges of about 400 yards, but rifled, small arms became common in the middle 1800s, and in the infantry's hands, their ranges were equal to or well-surpassed those of the best smoothbore artillery. With his rifled weapon, the infantry could now silence an artillery battery from beyond the range of canister. The range of a smoothbore muzzle-loader would always be short since the projectile had to be considerably smaller in diameter than the bore, and a tight fit was not possible. The field gun needed a longer range to protect itself and the rifled field artillery piece was the only answer. But the Europeans soon discovered it was not easy to muzzle-load a rifled gun; the projectile had to engage the rifling of the bore and had to be a tight fit. The answer to this problem was a breech-loading piece. Rifled breechloaders were developed in Italy and England in the 1850s, but all had problems with leaking gases and sometimes exploding when fired.

The U.S. War Department also authorized experimentation with rifled artillery and began rifling the bores of its bronze field pieces. One type, the James rifle, was able to throw 12-pound projectiles from a gun designed as a 6-pounder, but it was not adopted. Robert P. Parrot, at West Point, invented and built an iron, rifled, 3-inch muzzle-loader that could hurl a ten-pound shell just under two miles. Unfortunately, the gun crew couldn't see a target that far away. American artillerymen ignored these advances, all of which admittedly had drawbacks, and chose to stay with bronze, smoothbore muzzle-loaders. Peacetime deterioration plagued the U.S. artillery at this time. Several batteries were dismounted, and the remaining mounted batteries were dispersed west to frontier regions where disuse spelled their demise. Artillery training was episodic at best. Men were still rotated between field and foot duty, foot artillery being siege, garrison, coastal and rocket, up to 1858. In that year, an artillery school was established at Fort Monroe, Virginia, primarily for foot artillery, but in 1859, training in field artillery was added. The school did not meet expectations since the Army continued its policies of duty rotation and dispersing artillery units to the frontier to serve as infantry. On the eve of its Civil War, the United States had the best obsolete artillery pieces of the day and few trained crews to use them.

The American Civil War 1861–1865

This conflict, arguably the bloodiest in which Americans were ever involved, pitted 22 million people in 24 Northern (Union) states against 9 million people in 11 Southern (Confederate) states. Total Union strength had two million men under arms, compared with 750,000 under arms in the Confederacy. The South was an agrarian economy based on slavery, while the North was much more industrialized. The reasons for the war were many, but initially it revolved around state's rights, embraced by the southern Confederacy, versus a strong, centralized government of the Union northern states. Eventually, the problem of slavery, festering for many years, became a central issue. The South's economy was dependent on slavery, abolished in most nations by this time, and the North condemned the practice on moral grounds. Abraham Lincoln's election to the Presidency only added to the fears of the South — fears that this President would abolish slavery and destroy their way of life. Positions hardened and South Carolina seceded from the Union on December 20, 1860, followed in January 1861, by Mississippi, Florida, Alabama, Georgia, Louisiana and finally Texas on February 1, 1861. Four more states — Virginia, North Carolina, Tennessee and Arkansas — would secede in April. The central government in Washington could not tolerate secession and only an overt act of hostility on one side or the other was needed to trigger the war.

Captain Edmund C. Bainbridge's Battery A, 1st U.S. Artillery, at the Siege of Port Hudson in Louisiana in 1863. (Photo courtesy of the National Archives)

1861

The Union's Fort Sumter, in the harbor of Charleston, South Carolina, was fired on by Confederate artillery on April 12, 1861, marking the beginning of hostilities between the states. There was no declaration of war. The American Civil War was a modern war in many ways. Railroads transported masses of troops; both sides used trench warfare; balloons were used for troop and artillery observation; ironclad and steam warships battled each other and bombarded shore targets; and the course of the war was recorded by photography. Rifled artillery was only in the introduction stage at the beginning of the war — muzzle-loading smoothbores were the rule on both sides. Six- and 12-pounder guns and 12- 24- and 32-pounder howitzers were the artillery pieces available to both sides, all horsedrawn of course. Ammunition consisted of solid shot, canister, spherical case and explosive shells.

There were a few battles during 1861, as each side tested the capabilities of the other. On July 21, the Union's Major General Irwin McDowell started a march to Richmond, Virginia, the Confederate capital, with five divisions, including cavalry and about 50 artillery pieces of varying calibers; his army totaled about 30,000 men. McDowell met the Confederate's G.T. de Beauregard with 22,000 men at Manassas Junction, Virginia, near Bull Run Creek. The first shots of the morning were by the artillery of Union Colonel Ambrose Burnside's brigade against Confederate Colonel Nathan Evans. McDowell ordered massive artillery to the front and the pieces were moved to within 400 yards of the rebel lines to tear holes in the Confederate line. But the South's infantrymen were armed with rifled muskets, and at that close range, the North's artillery crews were felled en masse. Forced back at first by Union artillery and infantry charges, the South rallied, took the Union guns, and then lost them in a second Union charge.

In the afternoon, the South brought up reinforcements and with J.E.B. (Jeb) Stuart's cavalry and Jubal Early's infantry brigade and with artillery, attacked and turned the Union's right flank. Unequal to the task, the Union Army was forced to retreat from the field and the first battle of Bull Run (Manassas) went to the Confederacy. Neither side used artillery to its best advantage. Artillery could be used out of range of the rifled musket, but instead was used at close quarters where the opposing infantry could decimate its crews. Both sides were using small-unit Mexican War tactics in an era of massive armies and rifled weapons. McDowell was relieved by General George B. McClellan who, within eight months, completely reorganized his field artillery, increasing guns and ammunition per battery and the number of batteries per division. By the spring of 1862, the Union's artillery had grown to more than 500 guns and 12,500 men. The Confederacy reorganized its artillery, but its limited manufacturing facilities did not permit such extensive refitting of its batteries.

Two more Confederate victories in 1861 closed the year for a much-sobered Union Army. In August, Union troops were defeated at Wilson's Creek, Missouri. Missouri was a so-called border state, but now came under complete Confederate control. In October, a Union force was routed with 2,000 killed at the Battle of Ball's Bluff, Virginia. 1861 was a good year for the South.

East 1862

The second year of the war would go better for the North, although it would suffer defeats. On March 9, 1862, two ironclad ships, the Union's *Monitor* and the Confederate's *Virginia* (the former Union's *Merrimac*) fought to

"Field Artillerymen" by Joyce Kreafle

a draw off Norfolk, Virginia. Later in March, McClellan landed 100,000 men at the mouth of the York River (in the Peninsular Campaign), and marched to within six miles of Richmond by late May. The Confederate commander defending Richmond, Joseph E. Johnston, overran the Union's front line of infantry, in spite of a rain of shell and canister from McClellan's smoothbores. As the Union second line was about to be engulfed, massed Union artillery fire under divisional control stopped Johnston's charge and forced his men back. The South was unable to mass fire in a counterbattery exchange, in part because its artillery was attached to brigades rather than to divisions. Johnston was wounded in these battles and was replaced by Robert E. Lee as the new commander of the Army of Northern Virginia.

Lee, with 85,000 men, attacked McClellan and, in a series of some half-dozen major battles between June 21 and July 1, 1862 (Battle of the Seven Days), forced McClellan to eventually withdraw from the field. Before the Union Army withdrew, it was able to shatter Lee's attacks with massed artillery at Malvern Hill, east of Richmond. But the Confederates forced a Union Army to retreat. Lee reorganized his artillery in August, attaching batteries to the division that now enabled him to mass fire in time for the second Battle of Bull Run (Second Manassas).

On August 29, Rebel artillery in massed fire rained shell and canister on John Pope's attacking Union infantry and forced Pope to retreat. The next day Pope hit the Confederate center commanded by Brigadier General Thomas J. "Stonewall" Jackson and James Longstreet, and was decimated by fire from 48 Southern artillery pieces. A Confederate infantry attack then forced Pope to withdraw again. Lee's restructuring of his artillery was instrumental in winning the day for him, for now he was able to bring large numbers of guns together to mass fire on a single target. On the other hand, McClellan used his artillery often in single batteries only and then within range of the enemy's rifled musket fire. This was the same organization that plagued the Southerners defending Richmond two months earlier. Pope apparently had no time to learn from McClellan's recent Peninsular Campaign and repeated his mistake.

Both sides reorganized their artillery following the Second Bull Run. The Union assigned two or three batteries to each division. So did the Confederacy, but here Lee had a more centralized command. On September 17, 1862, a Confederate force was defeated by McClellan in the Battle of Antietam, Sharpsburg, Maryland. McClellan's corps commander, Joseph Hooker, with 90,000 men, launched an attack against Lee's Jackson, who commanded 50,000 troops. Union long-range, rifled artillery decimated the Confederate artillery early in the battle, but enough was left untouched for Jackson's artillery to stop the Union charge with canister. A rebel counterattack was then stopped by Union artillery and the Confederates did the same, a second time, to counterattacking Union infantry. Then McClellan's Ambrose Burnside, with reserve and division artillery support, hit Lee's right flank, which began to fold. The Confederates then brought up their reserve artillery and stopped the Union charge. As the day wore on, Lee began to suffer great losses and eventually had to withdraw to Virginia.

There were more than 23,000 killed and wounded on the two sides at Antietam, ranking it as one of the bloodiest one-day battles of the long war. Lee had learned to mass fire his artillery at critical times although his counterbattery fire (artillery fire against artillery), could not achieve a knockout punch due to lack of equipment. Offensively at Sharpsburg, the Union did not employ massed fire to support their infantry charges, so even with their superior rifled pieces and superior numbers, the North did not use them to full advantage. And its commanders did not learn the lesson, for no reorganization took place. Rifled pieces were being used more and more but when the chips were down in an enemy infantry charge, the defenders turned to firing canister from smoothbore cannons. The enormous destructive power of artillery was just at the threshold of being realized. Five days after this epic battle, President Lincoln announced the preliminary Emancipation Proclamation.

McClellan was replaced by Burnside after Antietam, and the new commander decided to march to Richmond from the north. On December 13, 1862, he was met by Lee's army at Fredricksburg, Virginia, about halfway between Washington and Richmond. Lee was entrenched with 260 field pieces looking down on Burnside's approaching army; the Union brought up 218 pieces to face the town. But there was no central command of the Union artillery and the pieces were employed piecemeal. Many batteries did not enter the fight at all. Again, the Union did not have the organization to mass fires in support of an infantry assault. Lee's artillery halted charge after Union charge led by Burnside's General George Meade against the Confederate positions on Marye's Hill. Suffering 12,000 casualties, Union forces were forced to retire from the field. Burnside asked to be relieved of command and he was replaced by Joseph "Fighting Joe" Hooker as commander of the Army of the Potomac.

West 1862

In western fighting, the Union's General Ulysses S. Grant captured Fort Henry, on the Cumberland River on February 6, and on the 16th, took Fort Donelson on the Tennessee River. Both forts were in Tennessee near the Kentucky border. Albert S. Johnston was forced to take his Confederate Army of 40,000 to Corinth, Mississippi, just beyond the Tennessee border, and Kentucky was saved for the Union. Grant, with 42,000 men, gave pursuit and the armies met at Shiloh (Pittsburg Landing) in Tennessee, 30 miles from Corinth. The ensuing battle raged through April 6 and 7. Albert Johnston was killed in the fight and Beauregard assumed command of the Confederate Army. Grant's army suffered 10,000 casualties and was shattered and forced to retreat after the first day's fighting. But, with 20,000 reinforcements from the Don Buell's Union army, Beauregard was forced to retreat south. Both sides used their artillery in an outmoded fashion, attaching the batteries to brigades, moving their cannons to within rifled musket range and suffering the consequences. The rough terrain precluded rapid deployment of the pieces on the battlefield. On the second day of this battle, Beauregard saw the wisdom of moving his artillery back out of musket range, but by then it was too late. On May 30, Grant marched into Corinth and a week later was in Memphis. On May 1, Captain David Farragut's naval squadron forced the surrender of New Orleans. The city, and much of southern Louisiana, was occupied by Union forces for the rest of the war.

The Confederate's Braxton Bragg invaded Kentucky in October, and was met at Perryville, Kentucky, on October 8 by Buell. The outcome was indecisive, but Bragg withdrew from Kentucky back to Murfreesboro, Tennessee. The North's William Rosecrans replaced Buell and the two armies met again at Murfreesboro (Stone's River) on December 31, in a three-day battle that went into the new year. Rosecrans' Army of the Cumberland was 47,000 strong. In one of the few instances of massed fire, the Union artillery stopped four separate Rebel infantry charges. The artillery, in fact, saved Rosecrans' army. After losing 9,000 of his 38,000 men, Bragg withdrew on January 2, 1863.

As 1862 ended, the Union had several victories to its credit, but in the rugged terrain of the West, artillery was of value only when it could mass fire on the defense. Longer-range rifled artillery was of little value since artillery was used in direct fire. The gun crews had to have their target in sight — usually less than 1,000 yards — and smoothbore tubes were sufficient for that. On New Year's Day, 1863, President Lincoln signed the Emancipation Proclamation, freeing the slaves.

East 1863–1864

Both sides now fully appreciated the value of artillery in a massed-fire role, and the necessity of centralizing command to achieve this end. Each side would appoint strong chiefs of artillery during the year and the Confederates grouped their batteries into battalions for more positive control. The Union curiously maintained its artillery control at the brigade level and initially its ability to mass fire was diluted. Between May 1 and 4, Joseph Hooker's Army of the Potomac was soundly outmaneuvered and beaten by Lee, Jackson and Jeb Stuart at Chancellorsville, Virginia. Hooker hoped to dislodge Lee from Fredericksburg on the Rappahonnock River by a flanking movement at Chancellorsville, west of Lee's main body. Lee saw it coming and attacked Hooker's flanking force on May 2, almost cutting the Union Army in two. Hooker's right was forced to pull back, but Lee's General Jackson, mistakenly wounded by his own troops on the second day of the battle, would later die of pneumonia. Hooker had no artillery chief at this time; his 400 guns had no overall commander and they did well to slow an overpowering Confederate charge on his right. Rebel batteries massed fire on Union infantry and broke up their formations, forcing a retreat. Then Stuart overran one Union battery after another and, after his right flank completely collapsed, Hooker saw no way to win. He withdrew his army in defeat on May 4. Because of this defeat, Hooker reorganized his artillery into brigades (same as the Confederate battalion) of four to six batteries each, assigned a brigade to each corps (two to the cavalry) and appointed strong chiefs of artillery with command powers. Hooker also appointed a strong Chief of Artillery, Major General Henry J. Hunt, to oversee it all. The Union Army was learning.

More has been written about the next battle in the East than any other of this War Between the States. It was the battle at Gettysburg, Pennsylvania, between July 1 and 3, 1863, that involved 165,000 men in the two armies. When

A Confederate-owned Napoleon gun used in the defense of Atlanta in 1864. Photograph by George N. Barnard. (Photo courtesy of the National Archives)

it was over, about 40,000 casualties on the two sides lay on the field of battle. Flushed with victories at Fredricksburg and Chancellorsville, Lee decided on a bold move along the Shenandoah Valley into Pennsylvania, hoping for a major, decisive battle with the Union and his Army of Northern Virginia. At this time, General George G. Meade had been given command of the Army of the Potomac in Lincoln's continuing search for a winning commander. Meade immediately marched his army northward, 90,000 strong, in pursuit of Lee's force of 75,000 and they converged on Gettysburg where an initial skirmish took place on July 1 between a foraging party of Confederate infantry and a scouting party of Union cavalry. The two forces deployed just west of Gettysburg along high ground called McPherson Ridge.

On this day, Confederate columns outflanked Meade's forces on this Ridge, and Union troops were forced back (eastward) to Seminary Ridge. Lee continued his attack and despite massed fire from the Union's 3-inch Ordnance rifles, Union forces were forced farther eastward to Cemetery Ridge, placing them due south of Gettysburg. Union reinforcements were ordered to Cemetery Ridge, for Meade now saw this line would define the major engagement between these two enormous armies. On the Union's right, and closest to Gettysburg, was Culp Hill and Cemetery Hill. Extending south was the three-mile-long Cemetery Ridge ending in two hills, Little Round Top and Round Top. Northern forces held all this high ground. Along this line, Meade positioned his artillery of 370 guns, now assigned to brigades and under command of chiefs of artillery in the field. This new organization was untested, but guaranteed that no battery would be idle in the coming battle.

Lee attacked again on July 2, opening with a barrage of some 80 pieces on the Union left; Lee had about 270 guns under his command and Lee's General Longstreet stormed the ground between the two Roundtops. Although he virtually destroyed a Union corps, he was stopped by massed fire of shrapnel (spherical case) and canister from Union artillery. Lee concurrently attacked the Union's right at Culp and Cemetery Hills, and was more successful, capturing several Union field pieces. But again, Union reserve artillery massed fire on these hills, and the Confederate attack was halted. On July 3, Lee ordered an attack on the Union center along Cemetery Ridge and opened with a barrage from 160 field pieces. Following a deafening artillery exchange with the North on Cemetery Ridge, Confederate General George E. Pickett charged up the slope with 15,000 men. Union artillery concentrated shell and canister fire on these troops, now in an open field be-

A 15-inch Rodman gun in Battery Rodgers, Alexandria, Virginia. (Photo courtesy of the National Archives)

fore the Union positions, and tore huge gaps in the even gray lines. Still, the gray army advanced and, in a vicious 20-minute battle within rifled musket range, was finally stopped and turned back. Union artillery meanwhile shelled Confederate reserves and prevented them from joining the fight. With the loss of more than 22,000 men killed and wounded at Gettysburg and now short on ammunition, Lee withdrew his battered army back into Virginia. Meade's army suffered more than 17,000 killed and wounded in the three-day carnage and was too exhausted to give pursuit.

In this massive engagement, regarded as a turning point in the War, Union artillery was employed with remarkable precision. Damaged pieces were moved to the rear, repaired if possible, resupplied with ammunition (and new crews, if necessary) and then awaited assignment or were immediately reassigned to the front. Pieces removed from the battle were replaced on the spot from the artillery reserves. Ammunition trains had been organized to guarantee no battery would run out of ammunition. Working to the Union advantage in the final battles at Gettysburg was a severe shortage of ammunition for the Confederate cannons. Other than this shortage, the South's artillery, organized along lines similar to the North's, acquitted itself well throughout this pivotal engagement. The war would continue for 22 months after Gettysburg, but in the East, the heavily wooded terrain and increasing tempo of battle would prevent artillery from playing a decisive part.

In May 1864, Grant took the 115,000-man Army of the Potomac south to an interior part of Virginia, west of Chancellorsville, to force a battle with Lee. The area was a mass of thick forest and undergrowth known as the "Wilderness." Lee marched 75,000 men westward to meet Grant, and on May 5 and 6, these two massive armies fought in this jungle-like area where artillery shells rained down on both sides, but cavalry was of little use. There were 28,000 dead and wounded on the two sides before Lee forced Grant to change the direction of his thrust. Grant turned eastward toward Richmond, Virginia, and the two armies met and fought again at Spotsylvania Court House, Virginia, from May 8 to May 12. Nineteen-thousand more men, blue and gray, died or were wounded. At Spotsylvania, Lee used his 3-inch Ordnance rifles to cut down Union charges but still the terrain permitted only occasional direct fire. Grant continued to push south toward the Confederate capital, and again, Lee stood his ground, this time at Cold Harbor only a few miles from Richmond. In a disastrous frontal attack on June 3, Grant charged his men directly into Lee's entrenched main body. Confederate canister and musket fire destroyed the attack, cutting down 6,000 Union troops in one hour.

Grant's losses in a month of fighting were a staggering 55,000 men, almost equivalent to Lee's remaining army. Grant was forced to change these senseless, massive frontal charges against a dug-in opponent. So Grant took his army around Lee, crossed the James River and marched toward Petersburg, south of Richmond. Lee reached Petersburg first, dug in his army and waited for Grant. Realizing his frontal attacks were useless, Grant began a siege of attrition that would last from June 20, 1864, to April 2, 1865, before Lee was finally forced to vacate the town. A week later Lee surrendered his battered army to Grant.

While the fighting in the "Wilderness" was taking place in May 1864, the Union's General William T. Sherman was driving the Confederate Army before him southward out of Tennessee and toward Atlanta, Georgia. In most of these actions, artillery contributed only marginally, except in defensive actions, due to the rough, wooded terrain. Sherman entered Atlanta on September 2, and on November 15, he began his notorious, 40-mile-wide slash-and-burn "march to the sea" to Savannah, which he entered on December 21, 1864.

West

1863–1864

On May 14, 1863, the Union's Sherman defeated a Confederate force in the Battle of Jackson, Mississippi, and Grant, marching south from Memphis, Tennessee, laid a six-week siege on Vicksburg, Mississippi, ending on July 4, 1863. The Union now controlled the Mississippi River to New Orleans and split the Confederacy into east and west halves. In September, the Confederate's Bragg trapped Rosecrans' army of 55,000 at Chickamauga, Georgia, and gave Rosecrans a severe mauling, forcing his retreat to Chattanooga, Tennessee. Again, the terrain prevented any massed firing of artillery by either side, and so artillery did not play a significant part in this battle. In November, Bragg moved north to contest Rosecrans for the control of Chattanooga. Grant now controlled all Union forces in the west and on November 23, 1863, brought together 40 field pieces that crumbled the Confederate's left at Lookout Mountain. Then massed fire on the Confederate's right forced Bragg back on that flank and finally, Grant hit the center and decimated both Confederate artillery and infantry. Bragg still had not learned the value of massed artillery fire and suffered from it.

On March 9, 1864, Grant was appointed to the command of all Union forces. On November 16, the Confederate's General John Hood invaded Tennessee and came up against the Union's Generals George H. Thomas and John M. Schofield at Franklin, Tennessee. Schofield commanded 30,000 men. The battle was joined on November 30, and it was not much of a contest this time. Hood lost 6,000 men and 11 Generals killed or wounded. On December 15, Thomas and Schofield caught up with Hood, again, at Nashville and in a two-day battle dealt him another deadly blow. Hood retreated into Mississippi. Again, artillery could not be used effectively in this rough terrain.

1865

The War was winding down by the winter of 1865. The South was all but destroyed and could not find materiel with which to wage war. Ammunition, clothing, food and shoes were all in short supply. Sherman turned north from Savannah and continued his march of devastation into the Carolinas. Charleston and Columbia, South Carolina, fell to him in February, and he reached Durham, North Carolina, in April. Richmond and Petersburg fell to the Union's northern armies on April 2, and Lee surrendered to Grant at Appomatox Court House, Virginia, on April 9, 1865. Lincoln was assassinated in Washington on April 14. By May, all Confederate armies in the east, south and west had surrendered to Union forces, the last on May 26, at Shreveport, Louisiana.

The Civil War did not see the replacement of the smoothbore muzzle-loaders by rifled, breech-loading artillery pieces, although these latter were available and were used by both sides. Technical problems with breech loading simply could not be solved, and cleaning a rifled bore took more time and was more difficult than a similar operation on a smoothbore tube. The rugged, smoothbore, muzzle-loading, 12-pounder Napoleon was the mainstay of the war. It was accurate enough for close-range direct fire, and light enough to be maneuverable on the battlefield. It could be fired rapidly and safely by an experienced crew and could be loaded with a variety of projectiles. In this war of massed armies, the artillery reserve became all important as a source of massed fire to stop infantry charges by thousands of troops. And in this war, the rifled musket of the infantry forced placing artillery out of musket range behind the attacking infantry. As a result, artillery used in the American Civil War became more a defensive weapon than an offensive one.

Artillery After the Civil War 1865–1898

Congress predictably reduced the size of the American Army after the Civil War. Artillery was reduced to ten batteries in two regiments of 12-pounder Napoleons (smoothbore muzzle-loaders) and 3-inch Ordnance rifles in separate batteries, one each per regiment. The ten batteries were scattered throughout the States, and each was manned by four officers and 76 enlisted men. Indian wars kept the army alert in the 1860s, although commanders rarely used artillery against the American Indian. The muzzle-loader/breechloader controversy continued in the U.S. Army into the 1870s without a hard decision between the two being reached. But not so in Europe. The Krupp works in Germany had solved the leakage problems with breech-loading guns by 1870, and European armies soon began replacing their smoothbore muzzle-loaders with new rifled breechloaders. These new steel guns had effective ranges of more than 4,000 yards compared to the smoothbore's maximum effective range of 1,000 yards with shot.

The next 30 or so years to the end of the century saw rapid strides in artillery design. Hydraulic spring-recoil mechanisms were perfected, and along with fixed ammunition (where the projectile and propellant were manufactured as a single unit to be loaded into the breech), a rapid firing of ten rounds per minute could be achieved in the 1870s. An experienced muzzle-loading crew could fire no more than three rounds per minute. By the last decade of the 19th century, the French had developed an excellent recoil mechanism that allowed an experienced crew to fire up to 30 fixed rounds per minute from a 75-mm (2.95 inches) rifled, breech-loading tube. Also by the 1890s, smokeless powder (a combination of nitric acid and cotton called nitrocellulose) was developed, which burned slower than black powder, with less pressure on the breech, but which exploded with nearly four times the force of the old black powder. As a result of these advances, artillery pieces could be made lighter and with longer tubes. This latter feature markedly increased the range of a projectile (in the case of the French 75-mm) to eight thousand yards. Finally, by the end of the century, the French had developed a shell explosive of nitrocellulose and picric acid they called melinite and whose explosion could be controlled precisely by the fuses of the day.

Direct fire of artillery was the rule up through the 1880s, where the target of shelling was visible to the gun commander, and he could quickly correct range and deflection by simple observation. With the newer long-range pieces it was now possible to shell targets over four miles away and invisible to the gun crews. This required developing a method of indirect fire where explosives could be laid on a target not visible. The answer was the forward observer who positioned himself to see the projectile explosions and then correct the elevation and deflection of the gun tube to bring it on target. Other pieces in a battery could then be adjusted to match the so-called base (corrected) piece. Immediately, communication with the forward observer, by initially unreliable field telephones, became of the utmost importance. Such communication was always tenuous since any mechanical device for communicating could be destroyed by enemy action.

The U.S. Army experimented with rifled breechloaders through the 1870s and early 1880s, initially converting the iron, 830-pound, 3-inch Ordnance rifle (of which there was an abundance from the Civil War) with a breech mechanism and reboring and rifling the tube to accommodate 3.2-inch ammunition. The M1885 gun proved superior to the smoothbore muzzle-loaders from which it was developed. Soon new tubes were cast of steel, so that by 1890, the Army had 100 of these new guns in service. The M1885 had an effective range of 6,600 yards but was not fitted with a recoil system and still used separate-loading black powder. But artillery was used on the frontier only as a defensive weapon against Indian charges on entrenched positions. Trained artillerymen were in the infantry, and trained gun crews almost disappeared. The Army did not recognize the field artillery as an independent branch of artillery and continued the confusing policy of rotating artillery officers between coastal and field batteries.

By 1896, the Army abolished its artillery reserve and placed all weapons in corps or divisions, recognizing that the longer range of the rifled pieces and their rapid firing capabilities could bring massed fires to the battlefield simply by redirecting the aim of relatively fixed guns. In 1892, the War Department established at Fort Riley, Kansas, the U.S. Cavalry and Light Artillery School, which trained professional artillerymen in their trade and taught cooperation with the horse cavalry. But on the eve of yet another war, America was still behind Europe in artillery technology. The M1885 gun, as noted above, still used black powder for the propellant and bursting charges, separated the propellant and the round, and was lacking a recoil mechanism. European armies had solved all these problems by the 1890s. The famous French 75-mm was fielded in 1897, and its advanced features, including a superior recoil absorbing mechanism, made all other artillery pieces obsolete.

The Spanish-American War 1898–1899

This brief war with Spain (actual combat lasted only about ten weeks) was a direct result of American imperialism and a hunger by certain "Expansionist" groups in the United States for overseas territory. Their target soon became Spanish-owned Cuba. Cuban nationalists, in 1895, rebelled on the island and were ruthlessly rounded up by the Spanish military and put in squalid concentration camps where many died. Some Americans were furious at this inhumane treatment and a hatred for Spain was fanned by some American newspapers. The American battleship *Maine* blew up in Havana harbor on February 15, 1898, and Spain was blamed for this disaster that killed 260 men of her complement. The true cause of the explosion was never determined. President William McKinley resisted attempts to intervene militarily, but Congress wanted to go to war. Spain was trying at this time to extricate itself from Cuba without seeming to give in to American pressure. On April 25, Congress declared a state of War had existed with Spain since April 21.

On the other side of the world, on May 1, Commodore George Dewey boldly steamed into Manila Harbor, Luzon Island, Philippines, with his six-ship squadron and destroyed a ten-ship Spanish fleet all lying at anchor. The army then occupied Manila, itself, on August 13, and Spanish ownership of the Philippines effectively ceased. In July, other Naval units had seized the Spanish Guam Island and unclaimed Wake Island in the Pacific. Congress then annexed Hawaii.

In and around Cuba, Spanish Admiral Pasqual Cervera sailed his six ships, four cruisers and two destroyers, into Santiago de Cuba harbor on Cuba's southeastern coast in May 1898, and was quickly blockaded there by a ten-ship American naval force. On June 22, 15,000 U.S. troops were landed at Daiquiri, just east of Santiago in Cuba, with the objective of taking the city. Major General William R. Shafter, Commanding General, V Corps, had in addition to his infantry, a provisional battery of four Gatling guns under First Lieutenant John H. Parker, a provisional field artillery battalion of four batteries of 3.2-inch field pieces under Major John H. Dillenback and two batteries of siege artillery. Shafter decided on an infantry assault on the village of El Caney, four miles northeast of Santiago, that sat atop a small hill and blocked the left approach to Santiago. As soon as El Caney was captured, his dismounted cavalry division and a second infantry division would make a frontal assault on the San Juan Heights complex, composed of San Juan Hill and Kettle Hill.

The attack on El Caney was launched on July 1, and did not go well. The artillery support was, in fact, ineffective since the four 3.2-inch field pieces used were fired at random and at unimportant targets. The infantry was forced to charge the defenses with little help from its artillery. In a bloody battle, the infantry finally drove the Spanish from each strongpoint and forced his retreat to Santiago. Artillery was moved to within 1,000 yards at one point and, finally, was effective in the bombardment of an enemy-held stone fort. Shafter, meanwhile, in the afternoon, decided to attack the San Juan Heights complex about two miles due east of Santiago, and began with a four-gun barrage of 3.2-inch shrapnel on the Spanish-occupied hills. The black powder of the American guns gave away their position, and Spanish shelling by their rapid-firing 3-inchers silenced them. Two hours later, the infantry attacked the hills, and the American guns awakened, again, to cover the charge.

Although enemy small arms and artillery fire made the infantry and dismounted cavalry positions almost untenable, the Americans continued their advance. Crawling along the ground, cavalry units, which included the First Voluntary Cavalry Regiment (Rough Riders) under Colonel Leonard Wood with Theodore Roosevelt as Lieutenant Colonel, seized Kettle Hill and caused the Spanish to flee to San Juan Hill. Scarcely had the infantry reached San Juan Hill when the First Voluntary Cavalry and the dismounted cavalry joined the attack on the hill. Together, American infantry and cavalry stormed Spanish defenses with support from Captain George S. Grimes' four-gun battery and a section of Gatling guns. As the Spanish retreated to his main line of defense, Major Dillenback moved Captain Clermont L. Best's and Captain Charles D. Parkhurst's reserve batteries, four guns each, forward to Kettle Hill to assist Grimes. By the time Best and Parkhurst had unlimbered, they could fire only one shot since American infantry and cavalry had already taken the crest of San Juan Hill.

From their positions on and near Kettle Hill, American artillerymen could not engage an enemy counterattack from the reverse side of the hill, if an attack came, because Americans employed direct-fire techniques, had flat-trajectory weapons and could not see the enemy. Therefore, Best wisely moved his battery to the top of Juan Hill to shell the enemy when it counterattacked. When the attack came, Spanish musketry and artillery drove Best's battery off the hill. Finally, after hard fighting, the Americans retook the hill and were able to place three field batteries in gun pits on line with the infantry.

This battle to take San Juan Hill cost the Americans 1,600 casualties. Shafter then moved his batteries to positions just east of Santiago and began a siege of the city. At this point, the Governor of Cuba ordered Admiral Cervera to run the U.S. Navy blockade and escape to sea. It was a disaster for the Spanish. In a one-sided battle, Cervera's squadron of four cruisers and two destroyers attempted to escape from Santiago harbor on the morning of Sunday, July 3, 1898. Most of the squadron was either sunk (the two destroyers) or run aground (three cruisers), and the fourth cruiser was chased out to sea and forced to surrender. Santiago surrendered on July 17, and neighboring Puerto Rico was occupied by American forces a week later. The peace treaty with Spain was signed in Paris on December 10, and ratified by Congress the following February. America was now a world power.

The fighting in Cuba revealed many inadequacies in the American artillery. Neither the artillery technology nor the tactics used by the Americans was up-to-date. The 3.2-inch guns used were breech-loading and rifled, but they had no recoil mechanisms, which slowed firing, and antiquated black powder was used as the propellant and the projectile explosive. The black powder billowed clouds of smoke, and the Spanish guns immediately had a position to shoot at. The Americans had no indirect-fire capability and had to move their guns forward to actually see the target. When the artillerymen moved up, they came within range of the enemy infantry's modern rifles and so could not hold forward positions. Their firing was often at about 2,500 yards from the target, and the results of their firing could not be seen. The result was that the artillery's support of infantry maneuvers was spotty at best. Although command of artillery operations was centralized and reasonably efficient, this command never took advantage of the capability to mass fires, even during the two-week siege of Santiago, where the guns were still employed piecemeal. All these deficiencies would be addressed over the next 19 years when America was drawn into the first of the modern wars.

Modernizing U.S. Artillery 1899–1914

The first attempt to update its artillery technology was only a half effort. Directly following the end of the Spanish-American War in 1899, a 3-inch gun with a short recoil system was designed and tested. It used separate propellant and projectile and had to be re-aimed after each firing. Tests eventually led to the M1902, a 3-inch rifled, breechloader with a long recoil system that reduced re-laying the weapon after firing, and which used the modern, fixed-ammunition and smokeless powder. The gun was accepted immediately and hailed as a state-of-the-art weapon. In rapid order there appeared the M1904 4.7-inch gun, the M1905 3.8-inch howitzer, the M1906 6-inch howitzer and the M1907 4.7-inch howitzer. These were excellent artillery weapons, but Congressional funding never allowed their production in quantities necessary to satisfy war needs.

The other major problem, the ancient, short-range, direct-fire method, was attacked early in the new century and very quickly indirect-fire methods were learned and adopted. Direct-fire methods died hard, however, and were used until the beginning of The Great War, World War I. Indirect fire, which took advantage of the 6,000-yard range of the new guns, required a whole new set of complicated tactics. A forward observer who could see the target now reported back firing corrections to the battery crew who were positioned too far back to see the target. In essence, the forward observer aimed the guns. At first, communication to the battery was by runner or signal flags, but the wired telephone came into use during the first decade of the 20th century. With this system, the observer only had to view the target, but not necessarily the battery. The battery could position itself in a woods or ravine, camouflaged. The observer was aided in World War I by sound ranging where banks of microphones could determine the position and distance of gun firing, and by flash ranging where the direction and distance of a muzzle flash of enemy artillery was reported back to a friendly battery, and the position of the enemy battery could then be determined. With the introduction of the airplane in World War I, the forward observer, at times, became airborne.

Unfortunately in this first decade, the training of U.S. artillery crews in these new methods was haphazard. In 1904, 1905 and 1907, various schools were authorized, but none came to fruition and a serious lack of trained artillerymen developed. In 1910, the War Department established the School of Fire for Field Artillery at Fort Sill near Lawton, Oklahoma. The following year, practical instruction began for commissioned and noncommissioned officers. The school produced artillery officers of admittedly poor, but improved, quality until 1916, when the school was closed due to the American border conflict with Mexico. Up to the eve of America's entry into World War I, the Army still did not have a corps of trained artillerymen or officers. Artillery tactics addressing the new long-range, rapid-firing pieces was in a constant state of flux in the United States.

Another major problem that surfaced in this first decade, but one that had been festering for some 80 years, was the overall organization of U.S. artillery. Coast and field artillery were still regarded by Congress as a single unit, and officers were still rotated between the two although tactics of coast and field artillery were entirely unrelated. Also, Congress recognized the battery as the highest level of artillery organization thus completely ignoring the necessity of massed firing, which such organization all but prevented. Beginning about 1905, prominent Army officers began to complain loudly about all these problems. These men promoted the concept of a regiment of three or more battalions as the basic field artillery unit and an increase in the total number of batteries, which would increase the ratio of guns per 1,000 soldiers. Finally, in January 1907, after much haranguing, Congress separated the coast and field artillery as independent branches and established battalions and regiments as the basic units of organization of artillery. This ended the policy of rotating officers between the two branches. Field artillery thus became an independent branch of the armed forces.

To ensure the infantry, cavalry and field artillery worked together, the War Department, by 1905, combined them in the division. The division consisted of three brigades of infantry (two or more regiments per brigade), one regiment of cavalry, nine batteries of field artillery organized as a provisional regiment of three battalions (total of thirty-six guns), one battalion of engineers, one signal company and four field hospitals. Each field artillery battery was composed of four officers and 160 enlisted men. The War Department, in 1908, replaced the division's provisional field artillery regiment with a field artillery brigade of two regiments of 48 guns and four battalions. The regimental staff consisted of nine officers and 45 men, and the battalion staff had three officers and five men. The battery was composed of five officers and one first sergeant, six sergeants, six corporals, two musicians, one guidon bearer, a driver for each pair of horses and had four guns, while a gun crew had two corporals and five privates.

In 1914, the War Department established four peacetime divisions, each with three infantry brigades and one artillery brigade with two regiments, plus attached engineer, medical and signal units. Each artillery regiment of two battalions had 24 3-inch guns plus eight howitzers (3.8-inch and 4.7-inch were in separate regiments). The Department also assigned one additional regiment of eight 4.7-inch guns and eight 6-inch howitzers to a field army. As war clouds darkened Europe in 1914, American artillery had finally entered a modern era of technology and organization. But a major problem existed in that there were virtually no trained crews to man the weapons.

At this point we will start to interweave U.S. Marine Corps artillery with U.S. Army artillery history, since the two share so many common events and developments. Much of early Marine Corps artillery history is the history of the 10th Regiment, the Marine's original artillery unit, formed in 1914.

Marines at Veracruz 1914

In a dispute with Mexico over the arrest of American sailors and refusing to recognize the revolutionary government there, U.S. President Woodrow Wilson dispatched U.S. Marines to the major Gulf of Mexico port of Veracruz to

"Marine Artillery at Veracruz" by Joyce Kreafle

Fifth Field Artillery Regiment howitzer sections in front of Signal Mountain at Fort Sill, Oklahoma, in 1912, when the Regiment was under the command of Colonel Granger Adams. Batteries of the Regiment were stationed at Fort Sill in the early 1900s until the time they were ordered to move to the Mexican border to guard against intrusions by guerrilla forces under Pancho Villa. (Photo courtesy of the Headquarters, 5th Field Artillery)

protect American interests there. Their orders were to take the Customs House and prevent the delivery of arms to revolutionary government units. Colonel John A. Lejeune formed an artillery battalion with 3-inch guns for this mission and manned it with 418 officers and men. The mission was achieved without firing a shot from these guns.

Marines in Haiti 1915

Haiti won its independence from France in 1804, but its history had been one of unstable governments almost since its founding. President Wilson sent the Marines into Haiti in 1915 to restore order following a coup attempt. The Marine artillery battalion, formed earlier by Colonel Lejeune, landed with 3-inch and 4.7-inch field guns, 14 in all, but the men fought as infantry rather than as artillerymen. U.S. forces withdrew from Haiti in 1934.

Battery A, 2nd Battalion, 218th Field Artillery, "Oldest battery west of the Mississippi River." Photo was taken at Seaside, Oregon, in 1916. They would practice by shooting at targets in the ocean. (Photo courtesy of Robert C. Fraser)

Battery A in Calexico, getting ready to chase Pancho Villa. (Photo courtesy of Robert C. Fraser)

Marines in the Dominican Republic 1916

Haiti's eastern neighbor, the Dominican Republic, won its independence from Haiti in 1844. Rebel political factions warred on the legal government in the early years of the 1900s and in 1916, U.S. Marines were sent to the Dominican Republic to restore order. The incursion resulted in the Marines' first use of its artillery in battle. The artillery battalion supported the Marines in their drive toward the city of Santiago and was given high marks in its close support bombardment of enemy positions the 4th Regiment's infantry assaulted and captured. The Marines occupied the Dominican Republic until 1924.

U.S. in World War I 1914–1918

The causes of World War I were many and deep-rooted but, in part, can be traced to a developing nationalism in Europe, a feeling that people who shared a common culture should live within their own borders with their own government, and not be ruled by foreign powers. Germany owned and occupied the French-speaking province of Alsace and much of Lorraine. Austria-Hungary controlled Slavic-speaking territories adjacent to Serbia, and also the Italian-speaking Trentino region and Trieste. Czechs and Slovaks also chafed under the rule of Austria-Hungary. People in the Balkan nations lived under Turkish rule, as did many different ethnic groups within the borders of Russia. All yearned for independence. With many of these vexing problems at the breaking point and with many nations of Europe armed to the teeth for battle, only a spark was needed to set off a conflagration. The spark that started World War I was the assassination, by pistol bullets, of Archduke Franz Ferdinand, heir to the throne of Austria-Hungary, and his wife Sophie, in Sarajevo, Bosnia (a province of Austria), by a Bosnian student, Gavrilo Princip. (A bomb attempt on Ferdinand's life earlier in the day by another would-be assassin had failed.) The date was June 28, 1914, a Sunday. All attempts at preserving the uneasy peace failed, and one month later to the day, Austria-Hungary declared war on Serbia. Within one week the major powers of Europe had aligned themselves with one side or the other — the Allies or the Central Powers — and the "Great War" was ignited.

The remaining five months of 1914 saw battles won and lost by both sides. Germany invaded Belgium and France, but her troops were stopped by the British and French at the Marne River, 23 miles from Paris, and Germany's hopes for a quick victory ended there. The Russian Army was defeated by Germany at Tannenberg, but the Russians, in turn, defeated the Austrians at Lemberg. There was little shifting of the battle lines on Germany's western front during 1915. One of the reasons was the evolution of static, trench warfare — 600 miles of trenches from Belgium to Switzerland. Both sides used poison gas, the Germans, first in April of 1915 (chlorine), and later in the year (December), phosgene. Although causing many casualties, gas attacks alone never caused either army to retreat from its position. 1916 saw many large drives on all fronts in an attempt to break the deadlock. Tanks were used for the first time in war by the British in this year. The battles of Verdun and the Somme are famous in military history as was the naval Battle of Jutland. The airplane came into its own in 1916, as a reconnaissance and bombing weapon, and was celebrated in the famous, but less useful, one-on-one "dogfights" by pursuit planes.

On April 6, 1917, the United States declared war on Germany, a response in large part to the many U.S. merchant ships that German submarines had been sinking, but also due to the discovery of secret German negotiations with Mexico, which offered Mexico large tracts of the American southwest *when* the Central Powers won the war. The United States was ill-prepared for war, as it would be, again, nearly 25 years later. America had a total of 200,000 men and 9,000 officers in its army in early 1917, and not a single division was organized or near full complement. The army would swell to 42 divisions and more than 4 million men by war's end. A Selective Service (draft) law was passed in May 1917, and American troops were in France by June 26 and in the trenches by October, fighting along the Somme River with British and French units. The first American artillery round was fired by Battery C, 6th Field Artillery Battalion, on October 23. This was also the year of the Russian revolution. When the Bolsheviks overthrew an interim government, Communist Russia signed a separate armistice with the Central Powers in December. These were dark days for the Allies as Germany now shifted its eastern-front armies into France in the hope of dealing the Allies a final death blow.

1917

General John Pershing was appointed Commander of the American Expeditionary Force (AEF) in June 1917. Pershing took to France the 1st Division made up of several already existing Army units. The area east of Verdun, in Lorraine, France, was the area of combat assigned the Americans. Although the AEF received combat training and experience at the front under British and French commanders beginning in October, they eventually gathered together in the spring of 1918 to form the First Army, solely under American (Pershing's) command. Pershing insisted on it. National Guard units quickly expanded to 50 regiments and provided four of every ten field artillery regiments that served in the AEF. At war's end there would be 234 field artillery regiments organized with a total complement of 460,000 officers and men. The artillery pieces used by U.S. artillerymen were almost all of foreign manufacture, including the famous M1897 French 75-mm field gun. The U.S. sent only about 100 guns to France during the war, and these were French 75-mm guns using plans supplied by the French. By comparison, the French alone fielded more than 3,800 guns.

"Fire Mission" by Joyce Kreafle

1918

The winter of 1917-18 was relatively quiet, but the Germans massed for a spring offensive in the hope of a quick victory before the full weight of American forces could be brought up to the line. They would be too late. By the middle of March 1918, there were 300,000 American soldiers in France. The number would swell to nearly 2 million by war's end in November. This final offensive began in March 1918, along a 50-mile front, and by the end of May, the German armies had reached the north bank of the Marne River, less than 40 miles from Paris. Two thousand Americans fought with the British near Amiens, Cantigny and Montdidier, on April 6, to stem the advance at the Somme River north of Paris. By the end of April, 500 U.S. troops were with the British at Ieper (Ypres), Belgium, and Armentieres, France, the latter just inland from Calais, and blocked the advance at these places. Early in June, U.S. troops, with the French, stemmed the German's farthest advance at Chateau-Thierry on the Marne River, and between June 9 and 15, 27,000 American troops, including American Marines, were instrumental in blocking the advance at nearby Belleau Wood.

From this time, the Central Powers coalition began to dissolve. The Allied forces now saw an end to this long conflict and launched counterattacks all along the front. July, August, September and October saw a steady rollback of the German armies in France. One by one the Central Powers capitulated until Germany was alone. Between July 18 and August 6, 270,000 U.S. troops were a major force in rolling back the Germans 20 miles at Chateau-Thierry and on August 8, the Americans with the British at Amiens began a counteroffensive that pushed the Germans back across the Somme to Cambrai. On August 19, 108,000 U.S. troops attacked with the British at Ypres and eventually forced the Germans to fall back to Ghent, Belgium.

Between September 12 and 16, over a half-million Americans in Pershing's First Army, preceded by a 3,000-gun, five-hour artillery barrage, attacked the long-standing German salient at St. Mihiel, France, on the Meuse River between Toul and Sedan. Two hundred sixty tanks, many of which broke down, were used in the drive, as well as 1,500 airplanes. Sixteen thousand German troops and 443 guns were captured in this drive that took the Allies to within a few miles of the German border. Finally, between September 16 and Armistice Day, more than one million U.S. troops drove north to take Sedan on the Meuse, capturing another 16,000 Germans and 468 artillery pieces. For the Allied offensive as a whole, on all fronts, more than 400,000 German troops were captured. Germany could hold out no longer, and on November 11, 1918, an Armistice brought this gigantic catastrophe to a close.

Loss of life on both sides was appalling in these decisive battles and reflects the final military dead in these most horrible of conflicts. For the Central Powers, 3,132,000 perished; for the Allies, 4,889,000 died, of which possibly 1,700,000 were Russian. Of America's final military force of 4.3 million men, 50,585 died in battle or from battle wounds and 205,690 were wounded. Total civilian dead on both sides is estimated at more than 6,640,000, many, perhaps most, from war-related diseases.

It has been estimated that 75 percent of the war dead were inflicted by artillery. Organization of artillery, generally, did not cause any significant problems on the battlefield in World War I, having become standardized by the European powers early in the conflict. Nor did the introduction of new weapons, other than the much used gas shell. Tactics, however, changed and evolved rapidly throughout the war. Massed fires and lengthy barrages reached unprecedented levels — preparation firing by 3,000 guns for nearly five hours at St. Mihiel — while other continuous bombardments lasted for more than two weeks. A dozen, very long ammunition trains steamed to the front almost every day in 1918. Sound- and flash-ranging were used to locate enemy batteries, and aerial observation was used, weather permitting. The Allies were able to field about twice the number of aircraft as did the Central Powers. Static, trench warfare allowed placing artillery pieces in pits for protection, and the rolling or "walking" barrage in front of an infantry advance was used by both sides. Calibers used got larger throughout the war and deadly gas-filled shells were used by both sides beginning in 1915 until the end of the war. Poison gas, however, was never significant in determining the outcome of any battle. Finally, there was some experimentation with truck-drawn, tractor-

Horses of the 5th Field Artillery Regiment are loaded on box cars in Toul, France, March 7, 1918. The unit was designated as the medium artillery battalion of the 1st Division and was equipped with French 155-mm Schneider howitzers. (Photo courtesy of the Headquarters, 5th Field Artillery)

drawn and caterpillar-drawn (tracked) heavy artillery, replacing the horse. In some cases, especially using tracked towing vehicles, it was successful, but came too late in the war to be effective. Some heavy pieces were, in fact, mounted on tracked vehicles — the beginnings of self-propelled artillery — but the combination proved too slow and unmaneuverable to be useful.

Divisional review of the 40th Division Field Artillery before California Governor William D. Stevens at Onda Vista, California. (Photo courtesy of the National Archives)

For the final three months of 1917 and the first half of 1918, American artillerymen were forced into the long-range artillery duels dictated by trench warfare. Corps artillery, with 48 155-mm howitzers in 12 batteries plus a battalion of mortars, took over the counterbattery mission of silencing the enemy's guns while division artillery with 48 75-mm guns in 12 batteries, plus 24 155-mm howitzers in 6 batteries and a battery of trench mortars, was tasked with rolling barrages to support infantry attacks. The final battles in the fall of 1918, always started with a massive artillery barrage, followed by infantry charges behind a rolling artillery barrage. This pattern was used week after week in the final assaults by the Allies against the Germans. These constant and aggressive attacks required rapid movement of artillery pieces on the battlefield, at which the U.S. troops excelled. Unobserved or map firing in which artillery fired on enemy targets based on his known or suspected position on map coordinates, was used by all forces. There were no forward observers to correct the fires. Often there was no help for this tactic, communications via telephone or radio was so tentative that the usefulness of a forward observer was often nullified. Also, on a land that was so completely laid waste, there was simply no safe place to position a forward observer.

Since they could not tie the infantry and artillery into a workable communication network, commanders had to depend upon elaborate plans and rigid schemes of barrages of unobserved fire and seldom employed observed fire. Corps artillery would fire a barrage on predetermined targets (hostile batteries, reserve locations, ammunition dumps or exposed flanks) that had been located on a map, while division artillery supported the infantry with rolling barrages. This intricate planning, which often excluded input from the infantry, minimized the need for either good communication or highly trained personnel and dominated artillery action. Although elaborate plans were designed to provide the needed support, these plans did not guarantee effective close support for the infantry when required. Battle conditions, terrain, weather and other factors generally prevented the infantry from following closely behind the rolling barrage, and these conditions often hampered counterbattery work.

Marines in France
1917–1918

The Marines went to France in June 1917, as infantry, to fight in the Great War. The Marine's first artillery regiment, the 10th Field Artillery Regiment, was ordered organized on January 11, 1918, and was formed on January 15, manned with the veterans of the Veracruz, Haiti and Dominican Republic campaigns. But the 10th Field Artillery Regiment and its 3-inch guns (76.2-mm) were not sent to France since Allied units there used the French 75-mm gun, and these guns were not given to the Marines.

Post-World War I Artillery Developments
1918–1939

Europe and the United States enjoyed a relative peace for nearly 21 years. As usual, America disarmed itself after World War I; after all, was not this "the war to end all wars?" Isolationism and a distancing from European entanglements dominated thinking in the United States for the next 17 years. But there was, in the years following the war, no lack of study panels tasked to look at the performance, or lack of performance, of artillery during the war. The so-called "Hero Board," headed by Brigadier General Andrew Hero, made many recommendations in 1919, including attaching light artillery to the division and heavier pieces to corps. The Board suggested that aerial observation be placed under control of the field artillery, and that light- and medium-weight pieces be motor-drawn in spite of the often poor quality of motor vehicles of the day. Hero did not see giving up the horse for moving light pieces, however. It recommended research on one new weapon, the 120-mm howitzer.

The Westervelt Board report of 1920, recommended sweeping changes in artillery, suggesting that every piece in the inventory be replaced and detailing the new calibers and number of guns, howitzers and mortars the Army should have, including a yet-to-be-developed 105-mm howitzer. Most important, America should manufacture all its own artillery and so end dependence on European supply. Experiments were initiated to test the three regiment (triangular) division compared to the World War I four-regiment (square) division of 28,000 men.

As a compromise, the War Department adopted a smaller (less than 20,000-man) square division armed with two 75-mm gun regiments and one howitzer regiment (105-mm when it became available). In the mid-1920s, the War Department began rearming the division with 75-mm guns, settling on a new 1923 model eventually to be known as the M1 75-mm gun. But there were so many of the old 75-mm (M1897) guns available that only a few of the M1s were ever purchased. Consequently, the M1897 was redesigned to withstand more powerful propellants and was designated the M2 75-mm gun. Its range now exceeded seven miles. By 1928, a satisfactory 105-mm howitzer (M1) was available; it was designed to be horse-drawn. Funds were limited, however, and so was production of the howitzer.

A crew and a 3-inch field gun in operation during the 1930s in Hawaii. (Photo courtesy of the Headquarters, 25th Infantry Division (Light) Artillery)

While all the searching for the ideal artillery combination was taking place in the 1920s and 1930s, there was a concurrent search for a suitable motor vehicle to tow the pieces. Budget cuts virtually curtailed the search. The War Department continued to view motor vehicles with suspicion, and with good reason in most cases — motor transport was, in fact, often unreliable. Self-propelled artillery, where the gun or howitzer was integrated with a motor and propulsion mechanism on the same chassis, met a similar fate. There were far-seeing men in the Army, however, General Harry G. Bishop, Chief of Field Artillery, 1930-1934, for one, who could see no reason why all artillery pieces should not be motor drawn, eliminating the horse completely. The horse, after all, had been vulnerable to World War I-type gas attacks. His insistence resulted in tests in the early 1930s, all of which were positive in favor of motor-drawn artillery pieces of all calibers and weights. During this time, little attention was paid to heavier pieces, the 155-mm and 240-mm howitzers. They were old and out-dated, but there were no funds to correct the problem.

By the early 1930s, however, the War Department began a gigantic modernization program for artillery. The Department ordered the motorizing of half the Army's artillery regiments in 1933, and in the next seven years, 75 percent of division artillery (75-mm guns and 155-mm howitzers) were motorized (truck-drawn) with new rubber-tired, pneumatic-wheeled carriages. Because of this new mobility, the triangular (three regiment) division, again, came up for review. Tests were successful, and by 1940, the War Department had adopted the triangular division. Three 75-mm gun battalions (36 guns in nine batteries) and one 155-mm howitzer battalion (12 pieces) were attached to the division. Eventually, the 105-mm howitzer would replace the 75-mm gun in division artillery.

While these changes were taking place, others in the Army were examining the problems associated with massed fire. By 1931, firing charts with coordinates were designed and the fires of a single battery directed to a specific coordinate. Firing elevations and directions were then calculated for all other batteries in the battalion to achieve massed fire. By 1932, experiments with a fire direction center were begun at the battalion level to compute all firing data for massed fires. Radios rather than field telephones were used for communication with the forward observers. This system evolved to be very fast and accurate. The fire direction center was a giant leap in the technology of massed firing, but, as with any change, it had

Members of the 13th Field Artillery, known as the "Clansman" Regiment, stand in a quadrangle in front of Schofield Barracks in Hawaii circa 1938. (Photo courtesy of the U.S. Army Air Corps)

its opponents. These voices were loud enough so the War Department at first did not promote the use of the fire direction center. By 1941, however, all the problems had been overcome and fire direction at battalion level was accepted throughout the Army.

The War Department, in 1941, adopted the fire direction center for the division. Using a system similar to battalion operations, the division fire direction center employed one gun from each battalion to adjust fire. By using the center, field artillerymen could mass fires from a battalion or division within minutes and put a maximum number of rounds on target with or without maps. The field artillery thus acquired the capability of delivering massed fires rapidly and shifting fire around the battlefield as necessary.

So, during these interwar years, the field artillery had evolved slowly in many areas. More fire power in the division, the fire direction center, continued experiments with the M2 105-mm howitzer (with a range of 12,000 yards), improved fuses in the late 1930s, more powerful propellants and motorization, all modernized the field artillery on the eve of World War II. In 1935, forward-seeing Major General Harry G. Bishop, Chief of Field Artillery from 1930 to 1934, characterized the artillery branch as the "King of Battle." But there was one soon-to-be-glaring weakness in America's artillery: it almost completely ignored developing any kind of antitank offensive tactics.

Marines in the Interwar Years 1918–1941

Demobilization after the war saw the 10th Regiment (Artillery) reduced to one battalion of a variable, but small, number of batteries. After the war, the Regiment's 3-inch guns were replaced by the French 75s and the 155-mm GPF. GPF was an abbreviation for

Grande Puissance Filloux (Great Power Filloux) named for its French designer, Colonel L. J. F. Filloux, in 1918. The Regiment continued to train through the 1920s and 1930s in both field exercises and amphibious techniques. In 1927, skirmishes between two military leaders near Shanghi, China, were seen as threatening American interests and citizens in the area. Consequently, the 10th Field Artillery Regiment deployed to Tientsin, China, with the 3rd Marine Brigade. It was designed as a show of force, and it had its intended calming effect. The Marines were home after a year of duty.

The term "Regiment" in the Marine Corps was dropped by July 1930, and replaced by the term "Marines," so the 10th Regiment became the 10th Marines. Also, about this time, the Marine Corps adopted the Army's 75-mm mountain piece as a pack howitzer. It weighed about 1,300 pounds and was capable of being disassembled and transported by six mules. The remainder of the 1930s were relatively quiet and given over to training as much as the economics of the Great Depression would permit.

World War II
1939–1945

World War II began in Europe on September 1, 1939, when German dictator Adolf Hitler's troops and armor smashed across the Polish border and the German Air Force swept clear the skies over Poland of enemy aircraft. The appeasement policies of France and Great Britain had failed to preserve the peace in Europe. On September 3, these two allies of former battles declared war on Germany. By early October, Poland had been crushed and Germany would eventually turn westward to overrun the Low Countries and France. With the onset of war in Europe, the United States was alerted that its peaceful, isolated life in North America would soon be changed. The Roosevelt government in Washington began selling war materials to the Allies in Europe, and in March 1941, President Franklin Roosevelt signed a "lend-lease" bill under which war materials could be shipped to the Allies as a matter of American policy and paid for later, he hoped. On July 10, 1941, there was established at Fort Sill, Oklahoma, the Artillery Officer Candidate School, a 13-week training program for warrant officers and enlisted men aspiring to become second lieutenants in artillery. The course was lengthened to 17 weeks in 1943. The School would stay in operation until 1946 and would commission 26,209 second lieutenants.

Following the fall of France, Hitler turned his attention to the invasion of the British Isles. July 1940 saw the beginning of an attempt by the Luftwaffe (German Air Force) to destroy the RAF (Royal Air Force), neutralize the British navy and reduce English cities and towns to rubble by bombing. This "Battle of Britain" would last into the spring of 1941, but the RAF had, in effect, broken the back of the Luftwaffe by October 1940, and Hitler was forced to postpone indefinitely thoughts of a cross-channel invasion. At a cost of 537 brave fighter pilots and 1,020 aircraft, the RAF had saved England. These attacks on English cities ended in May 1941, when the Luftwaffe was withdrawn for use in Hitler's invasion of Russia. On June 22, 1941, while the British were engaged in seesaw battles with German armor in North Africa, Germany launched 100 infantry and armored divisions eastward across the Russian border. The war on Russian soil would last into 1945, but the attack, although initially highly successful, would eventually fail.

Artillery tactics used by the Germans in their new "Blitzkrieg" (lightning war) were adapted to the high-speed mode of tank attacks and overwhelming air superiority. Massed fires first silenced enemy batteries in the area of the main attack. A first wave of tanks, not infantry, then cut through the first line of enemy infantry defense and radioed the position of enemy strongpoints to their artillery, which opened fire on the enemy. Following waves of tanks, infantry and artillery were brought forward to neutralize antitank and missed positions and to repeat the process. This required massed fires only in the initial attack and breakthrough. During these initial attacks, the Luftwaffe bombed supply depots, roads, railroads and reserve troop concentrations. In Germany's attack into the Low Countries and France, armor and mechanized infantry moved so fast that artillery units could not keep up, and artillery played only a small part in the fall of Belgium, Holland and France.

America Enters the War

The day was a Sunday; the date was December 7, 1941; the local time was 0751 hours. From six aircraft carriers 235 miles north of Hawaii, 183 Japanese fighters, torpedo bombers, medium altitude bombers and dive bombers struck the U.S. Navy installation at Pearl Harbor,

Gun display, D Battery, 13th Artillery, at Schofield Barracks in Hawaii in 1940. (Photo courtesy of the Headquarters, 25th Infantry Division (Light) Artillery)

Gun Section, D Battery, 13th Artillery, at Schofield Barracks in 1940. (Photo courtesy of the Headquarters, 25th Infantry Division (Light) Artillery)

Honolulu, Hawaii, and bombed and strafed several Army airfields on the island. At 0855 hours, a second wave of 167 planes renewed the attack on the Pacific Fleet anchored in the Harbor and again strafed Army airfields on the way out from their primary targets. By 1000 hours, all the Japanese planes were on their way back to their carriers. They had lost 29 planes and 64 of their own pilots and air crewmen to ground fire and Army Air Corps P-40 fighter pilots, but had destroyed or badly damaged about 300 American planes, mostly on the ground. Pearl Harbor, itself, was total chaos: three battleships totally destroyed and five more badly damaged; three light cruisers, three destroyers and seven smaller vessels sunk or badly damaged. There were no aircraft carriers in port at the time of the attack. All of the ships, except the three destroyed battleships, would be repaired to fight again. The worst statistic was the American casualty figure: 3,581, of which 2,403 were killed in action. Eventually, all the carriers, battleships and cruisers of the Japanese attack fleet — eleven ships in all — would be hunted down and destroyed. On December 8, the U.S. Congress declared war on Japan and a few days later on Germany and Italy.

The war in Europe showed America unprepared for the new kind of lightning war. Its first reaction was to build a self-propelled weapon for battlefield use. The weapon chosen was the M2 105-mm howitzer, which was mounted on a fully tracked medium tank chassis with a machinegun turret and was designated the M7 105-mm self-propelled howitzer. The pulpit-like turret gave the M7 the name "Priest" or "Sexton." It was makeshift but it worked, and the first production was shipped to the British fighting in North Africa early in 1942. M2 105s were soon mounted on half-tracks. The War Department later ordered all artillery pieces be adapted for truck towing. The pieces involved were many: all 75-mm guns (M1897, M1916, M1917 and M2), the M2 105-mm howitzer, the M1917 and M1918 155-mm howitzers, M1 155-mm gun, the M1 8-inch

"Armored Field Artillery – WWII" by Joyce Kreafle

howitzer and the M1918 240-mm howitzer, the M1 4.5-inch gun and the M2 8-inch gun. The M12 155-mm gun and the M7 105-mm howitzer were self-propelled. During Army maneuvers in 1941, light aircraft proved themselves superior in the air observation role, flying thousands of missions without mishap. By June 1942, artillery had its organic air observers, that is, aircraft dedicated solely to artillery needs and controlled by artillery. Each battalion was assigned two observation planes with pilots and mechanics. But still, field artillery refused a role in antitank and antiaircraft operations; the former was the mission of tank destroyer units with both towed and self-propelled weapons, and the latter was the responsibility of the coast artillery.

1942 was a year of building for the United States. There were few successes to bolster American spirits until near the middle of the year. The so-called Axis Powers — Germany, Italy and their allies — reached their zenith in Europe and Africa by September 1942. Then the Allies began to strike back. Rommel's Panzer divisions were halted at El Alamein by British General Bernard Montgomery's Eighth Army, and in four months were pushed back 1,100 miles westward to the Tunisian border. In November 1942, the Russians counterattacked at Stalingrad, the beginning of the rollback of the German hordes in Eastern Europe. Also in November, American and British troops landed in Algeria and Morocco, and Axis resistance, primarily Vichy French who took their orders from the Germans, was overcome in three days. In the Pacific in 1942, the Japanese were overrunning one Asian nation after another. In May, the U.S. Navy blocked Japanese expansion southward toward Australia in the Battle of the Coral Sea. In June, the Navy's weak but growing carrier force decimated a Japanese carrier fleet near Midway Island to block Japanese expansion eastward toward Hawaii. In August, the U.S. Marines landed on Guadalcanal in the Solomon Islands to begin the long and bloody westward roll-back of Japanese ground forces.

The War in Europe North Africa

American artillery was used in Morocco and Algeria following the landings there on November 8, 1942. The 175th Field Artillery Regiment landed at Algiers with the 168th Infantry Regiment, and they were on the heights commanding the city by the afternoon. Algiers surrendered on November 10. The American operation near Casablanca in Morocco was primarily an infantry/tank campaign. Casablanca fell on November 11. American dead for the Algeria-Morocco campaigns were 556. The real test came in the late winter and early spring of 1943 in Tunisia. The British Eighth Army had pursued the retreating German tank divisions from Egypt all the way to Tunisia and by February 1943, Germany's desert tank divisions were being forced into southern Tunisia, from which there would be no escape. Rommel had flown to Germany for medical treatment, but his tank commanders in Tunisia were not about to give up. They would fight to the end, now against American and Free French forces in addition to British. The Germans holding Tunisia had established a north-south defense line along a line of hills called the Eastern Dorsale that stretched from the Mediterranean coast, west of Bizerta, southward to a point about 20 miles south of Gafsa, an important road junction. The defense line was about 250 miles long. American troops were pouring into Tunisia by January 1943 to bolster British and French units holding the line against the Germans in central Tunisia. American artillery was about to be tested, and initially, it would not fare well.

On February 7, Colonel Thomas E. Drake (34th Division) was ordered to move his Regiment, the 168th Infantry less the 1st Battalion, to a cut through the Eastern Dorsale called Faid Pass. Through it ran the main road from coastal Sfax to Sbeitla, 110 miles inland. Colonel Drake was now placed in command of all infantry around Faid which, in addition to his own regiment, included a unit of the 109th Engineers, a tank destroyer unit from the 9th Infantry, a cannon company from the 39th Infantry and the 175th Field Artillery. There also arrived at this time 450 replacements, many of whom had never fired a rifle. They arrived without entrenchment tools, without arms and were weighed down with barracks bags filled with generally useless items. They were anything but infantrymen, which was the greatest need.

On the morning of February 14, more than 80 Panzer tanks attacked from the north and east. A small group of American tanks went out to meet this force and were immediately overwhelmed, losing seven tanks and several antitank guns. The terrain was mostly open and the forward units of our infantry were literally crushed in the path of the oncoming enemy armor. The air was filled with Stuka dive-bombers that dropped their lethal loads and then made low-level strafing runs on the infantry who could find no cover. The enemy was now encircling the 168th, and Colonel Drake ordered his troops to take a hill about a mile away as a last-ditch defensive position. The last telephone message received by the Colonel before the wire was cut said that all artillery support was being withdrawn, but that he should hold fast. The 109th Engineers quickly mined the approaches to the hill, and for the next two days, the best of the German infantry and tanks attacked the 950 men holding the hill, one-third of whom were unarmed. By February 16, all of the artillery pieces of the 175th Field Artillery were silenced. Colonel Drake pleaded for air support, reinforcements, ammunition and rations; none of these arrived. The situation was now desperate; the men hadn't eaten for three days, ammunition was running out and the Regiment was systematically being cut to pieces by German 88-mm guns and the Stukas. The 2nd and 3rd Battalions, 168th, were now ordered to withdraw westward as best they could. Colonel Drake led the 2nd Battalion in its retreat. All serviceable equipment was destroyed or disabled. The battered men of the 2nd Battalion covered a distance of 25 miles on the night of February 16 over open ground, without cover. The German Panzers, having broken through all defenses, caught some of these men and many other small groups of stragglers that were trying to escape through enemy positions. Colonel Drake was forced to surrender; he had nothing left to fight with. In small groups, the remainder of the 2nd Battalion that had split off from Drake's group fought its way westward through enemy positions in a continuing series of fire fights. Remnants of the battalions reached Sbeitla, 40 miles from Faid Pass. Of 800 men, 432 made it back to friendly lines.

The 3rd Battalion was the last to withdraw from the massacre at Faid Pass. For four days, scattered units of this battalion fought through enemy positions, but they were cut to shreds by the German tanks that roamed at will over the countryside. The entire battalion simply disappeared, save one officer and two infantrymen who staggered in on February 20. The Faid Pass debacle was the first and last time the 168th was ever forced to retreat before the enemy in the entire war. The losses to the 34th Division were staggering: 1,628 killed, wounded, missing or captured at Faid Pass. Losses in other units brought the number to 2,500. One hundred tanks, 280 other vehicles and 30 field pieces were destroyed or abandoned by the 168th at Faid.

With the breakthrough at Faid, the enemy tank force now thrust north toward Sbeitla and west toward the Kasserine Pass, about 40 miles from Faid. At this time, the other two regiments of the 34th Division (133rd and 135th Infantry) were in position at Pichon, about 40 miles north of Faid. On February 17, the 21st Panzer Division, breaking through Faid, hit the 3rd Battalion, 133rd, which had dug in, anticipating the attack. A rain of fire stopped the German attack, destroying one tank and two other armored vehicles, forcing the enemy to withdraw. The 133rd and 135th Regiments were then ordered to withdraw westward to Sbiba, about 30 miles, fearing an encirclement from the south by another column of enemy tanks. The 30 miles was a hit-and-run gun battle between the two regiments and the enemy tank thrust. On February 18, the two Regiments dug in near Sbiba and soon took the brunt of another tank attack. This time American artillery opened on the tanks and infantry with murderous effect, repelling the attack. But this position became untenable as it was reported German tanks were pouring through the Kasserine Pass to the southwest in an attempt to encircle the American position. On February 22, 9th Division Artillery, in position between Sbiba and Thala, struck a German tank column with 75-mm guns and 105-mm and 155-mm howitzers, with 36 other British pieces, 84 pieces in all. The enemy tank thrust was stopped and turned back. At the same time, 1st Division Artillery caught another enemy column of 38 tanks exiting Kasserine Pass to the southwest and forced it, also, to turn back. In these actions, radio-equipped observers attached to infantry or armor units were sent aloft in organic spotter aircraft, and with centralized command, field artillerymen repeatedly massed fire on German positions and targets of opportunity. On February 23, reconnaissance showed that the Panzer Divisions had withdrawn eastward from Kasserine toward their own lines. The westward thrust of the Panzer Divisions in Tunisia ended. At this time, the British Eighth Army was driving Rommel's desert divisions into southern Tunisia. Following the disastrous Faid attack, American artillery had acquitted itself well. From Dastrup (p. 211):

For the most part, commanders favorably commented about the field artillery's performance during the drive to push the Axis Powers out of North Africa. Reflecting on the effectiveness of his field artillery, Major General Manton Eddy, Commander, 9th Division, recorded, "One Nazi who had served on almost every German front said that the American artillery fire was the most deadly that he had experienced." After driving the Axis out of North Africa, Lieutenant General [Omar] Bradley, Commanding General, II Corps, commented, "....the American field artillery technique of massed fires was a major contributing factor toward the early and successful conclusion of the operation .... Specifically, the development of a centralized battalion fire control by Major General (then Major) Orlando C. "Pinkie" Ward, was the major reason for the splendid performance of our field artillery in World War II and earned great respect among allies and enemies as well.

Given the opportunity, the Army's field artillery demonstrated the impact of the fire direction center, ground and organic air observation and motorized light artillery. In most cases, observed fire proved to be more effective than unobserved fire because the hills and ridges in Tunisia provided excellent positions for observation and because each field artillery battalion had at least ten observers to adjust fire through the fire direction center. Addressing the center, General Eddy said that it united observers and battalions into an effective network. In one instance, a forward observer on the northern flank of Eddy's division sector conducted fire for a battalion of 155-mm howitzers on the southern flank.

For three months following the Kasserine Pass battles, Allied units relentlessly attacked the Germans in Tunisia, always pushing north and east toward the sea. Bizerta fell to American and French units on May 7, and Tunis fell to units of the British Eighth Army on the same day. On May 15, 1943, General Jürgen Von Arnim, commanding all Axis forces in North Africa in Rommel's absence, surrendered to the Allies. Nearly a quarter-million German and Italian prisoners were taken, many driving themselves to or simply walking into the Prisoner Of War camps, in addition to thousands of tons of war equipment. Allied artillery in most cases, functioned to near perfection throughout this final phase of the battle for Tunisia. Both self-propelled (M7) and towed artillery maneuvered quickly and surely around the battlefield, wherever they were needed. Their crews in these months became expert at their jobs.

The American II Corps' artillery in the Tunisian fighting, fired some 23,000 rounds per day. Every aspect of the U.S. modern artillery organization, training, tactics and equipment had been put to the ultimate test — combat — and all had passed with high marks.

Sicily

The next target on the Allies' "hit list" was the island of Sicily, east of Tunisia, which was used as a springboard for the invasion of Italy. This 38-day campaign was to be a fast-moving, infantry/tank-assault operation in which artillery played a lesser role. The assault forces left Bizerta on July 6, picked up other elements of the huge convoy at Tunis, and by July 9, 1943, were off Malta, 250 airline miles east of Tunis, and only 75 miles south of the planned assault beaches on the southern coast of Sicily. This was a gigantic amphibious operation — the biggest in military history — to be overshadowed only by the Normandy landings in France, some 11 months away. Three thousand ships of all kinds carried 160,000 men, 600 tanks and 14,000 other vehicles to the beaches. It was a joint American, British and Canadian operation, with British General Harold Alexander in overall field command under American General Dwight D. Eisenhower as Supreme Commander. Two Armies, the British Eighth Army under General Bernard Montgomery, and the American Seventh Army under General George S. Patton were to assault the southeast coast of Sicily; Montgomery on the right from Cape Murro di Porco around to Pozzallo, and Patton on the left from Cape Scaramia to Licata. The British would land four infantry divisions, part of an airborne division, an infantry brigade, two armored brigades and three commando teams. Their reserves were spread widely over North Africa and the Near East. In Patton's Seventh Army was the yet untried 45th (Oklahoma) Infantry Division, fresh from the States, the 1st and 3rd Infantry Divisions, the 2nd Armored Division and part of the 82nd Airborne Division. In reserve in North Africa was the American 9th Infantry Division and the remainder of the 82nd Airborne Division. Available to Alexander were 1,800 artillery pieces among the two Armies.

The enemy strength on Sicily was reported to be four Italian infantry divisions, six Italian coastal defense divisions (judged to be poorly trained) and two German infantry divisions, both of high quality and one of them armored. The Allies were able to put up 4,000 operational first-line aircraft against an estimated 1,800 Axis aircraft from Sicily, Italy, Sardinia and Southern France. Allied air bombardment began on July 3, and after one week of round-the-clock attacks, the German Luftwaffe was virtually eliminated from the impending battle zone.

The 3rd Infantry Division and part of the 2nd Armored Division arrived off Licata about 0300 hours, July 10, 1943. The infantry, reinforced by the 3rd Ranger Battalion, was put ashore immediately to clear out and hold a beachhead. At dawn, a battalion of Sherman M4 tanks was put ashore to enlarge and consolidate the infantry's beachhead. Resistance was light and some additional units of the 2nd Armored Division began to come ashore shortly after dawn. The small enemy garrison at Licata surrendered with little resistance, and the tankers immediately captured the airfield, again with minimal resistance. Now, they began drawing fire from the nearby hills, and the tanks deployed to attack from the front and the flank. In a brief search-and-destroy battle, the tankers destroyed six enemy 75-mm artillery pieces and a number of pillboxes with their gun emplacements.

On July 11, the Germans decided to strike the landing force. As dawn came, the 1st Division infantry troops cautiously pressed inland, expanding the beachhead. A platoon of the 82nd Reconnaissance Battalion probed eastward on the right flank of the 1st Infantry, looking for German armor. They found an advanced party of enemy tanks, called in a company of the 67th Armored Regiment (2nd Armored Division), and the enemy tanks were driven off. But this small enemy force was only a feint, for elements of two German divisions were forming up in a valley north of Gela and at 0800 on July 11, about 40 enemy tanks headed for the beachhead. These German divisions had been forced-marched from the western half of the island where they had been deceived into thinking the main landings would take place. They were short of artillery and infantry support, but had plenty of fuel and ammunition. The attacking enemy tanks were approaching from the northeast toward Gela and the beachhead on the Gela-Niscemi road when they ran into a battalion of the 1st Infantry's 16th Regiment on the American right flank. In a wild firefight lasting several hours, about 50 men with one antitank gun, one bazooka and later in the day a tank destroyer, drove off the enemy armor. The Germans were unable to turn the 1st Division's right flank.

On the morning of July 12, the 16th Infantry Regiment was ordered to push northeastward toward Niscemi, about eight miles away. Suddenly, in a master stroke, some 40 German tanks skirted around the left flank of the 16th, closed in to its rear, and made a dash for the beachhead. It was a brilliant, tactical move on the part of the German commander. If he could push the beach party back into the sea, he could then turn on the isolated American infantry. It almost worked. The 2nd Armored still had only a few of its tanks ashore at this time, and the situation became grim. A platoon of four tanks, led by Lieutenant James White, was ordered to straddle the road leading eastward out of Gela toward Vittoria and to do what they could to stop the German column. American artillery ashore was given coordinates of the counterattacking enemy tanks and began sending shells into the German force. As the enemy column appeared, White's four tanks commenced firing. Artillery, mortar fire and naval gunfire rained down on the enemy column. Then an M2 105-mm towed howitzer arrived to add to the tank fire. Three of White's tanks, having expended all their ammunition in the wild melee, pulled back to rearm. Seeing this retreat, the German tanks advanced to finish off the remaining tank. White and the howitzer opened up, destroyed one enemy tank and forced withdrawal of the others. Then a 105-mm M-7 self-propelled howitzer arrived, and shortly thereafter, newly landed tanks of the 2nd Armored Division joined the battle. This proved too much for the German tanks (the Hermann Goering Division). They broke off the engagement and withdrew to the inland hills. Fifteen enemy tanks were destroyed in this brief, but fierce, fight at a cost of three American wounded. The beachhead was saved. Thus it went for the additional 36 days it took to take Sicily. Artillery had a hard job keeping up with Patton's hell-bent-for-leather style of warfare, but when they were needed, they were there to add their fires to those of Patton's tanks. By August 17, American troops were in Messina, and the enemy had been cleared from Sicily. Two years later, when the War in Europe was over, Patton was quoted as saying, "I do not have to tell you who won the War. You know. The artillery did."

Italy

On the night of September 8, the announcement came that the Italian government had fallen and that all Italian troops were to cease hostilities. The Italian fleet at Taranto surrendered and British forces went ashore unopposed. The Germans immediately replaced all Italian forces with their own. The announcement that Italian troops were to stop fighting sent a wave of optimism through the assaulting troops of the 36th and 45th Divisions, already off Paestum, about 20 miles southeast of

An artillery crew of the 50th Field Artillery Battalion rushes to prepare a 105-mm howitzer for firing position while on maneuvers at Westdown Fields in Tidworth Wilts, England, October 2, 1943. (Photo courtesy of the National Archives)

Sergeant Richard J. Clements gives orders to fire a 105-mm howitzer as a gun crew from Battery C, 10th Field Artillery, 3rd Infantry Division, puts shells in the howitzer in Italy, October 12, 1943. (Photo courtesy of the National Archives)

Paratroopers of a 75-mm howitzer section, Battery C, 377th Field Artillery, 101st Airborne, put on their parachutes before loading into a C-47 cargo plane in Newburg, England, March 10, 1944. (Photo courtesy of the National Archives)

Men of the 48th Field Artillery Battalion clean and work on their pieces at the camp in Mourne near Kilkeal County in Ireland, March 13, 1944. (Photo courtesy of the National Archives)

Salerno in the 450 ships of the invasion armada. The Germans, however, had no thoughts of allowing the Allies a foothold in their Fortress Europe.

At 0100 on September 9, the men of the 36th Division scrambled down the ship-side nets and into amphibious trucks and headed for the beach. As the first boat scraped the shore, all hell broke loose. Flares lighted the sky and German artillery opened up on the beach. Two regiments of the 36th Division made it ashore and immediately rushed up the beach to the plains beyond, toward low hills faintly visible in the darkness. Once these infantrymen were ashore, the men of the 151st Field Artillery were lowered into their landing craft and headed for shore. One of their 105-mm howitzers was lost in a ramming accident, but all men were saved. When the 151st reached the beach, it became apparent that the landing was in jeopardy; the two regiments of the 36th were taking an unmerciful pounding from artillery and Panzer tanks brought up to repel the invasion. The 36th Division's artillery, including the 151st, quickly set up its positions and began pouring a rain of fire into enemy gun positions, enemy troops and advancing Mark IV tanks. The situation became critical; the infantry was being threatened by superior forces, and it appeared for a while that they, and the entire invading force, would be forced off the beachhead and back into the sea. It was at this moment of doubt when the 151st swung into action. Units of the division had penetrated to the foot of Monte Soprano, the dominant feature of the area, about seven miles inland from the beach. They were in danger of being cut off and encircled by German tanks. Seven Mark IVs attacked these units and the 151st leveled its guns and destroyed two of them. The remainder retreated. Tank groups attacked them repeatedly throughout the morning. The 151st beat off every attack. Between 1030 and 1330, 14 Panzer tanks assaulted the position of the 151st; five of these tanks received direct hits by our batteries and were destroyed. Each time, the remaining tank commanders of the group thought better of taking their chances against the 151st Field Artillery's fire, and withdrew. In one incident, one battery was overrun by a tank attack; the gunners withdrew, deployed as infantry and, with rifles and grenades, took back their position and resumed firing.

For four grueling days, the Germans pounded the beachhead and the plains beyond, but by then the landing was secure. On September 15, the German forces showed signs of weakening and started a general withdrawal northeastward. Of the 151st's outstanding performance in this difficult operation, the Chief of Staff of the 36th Division said: "Had it not been for the arrival of the 151st Field Artillery Battalion 30 minutes before the first counterattack by German tanks, it is probable that the beachhead would have been destroyed." To comprehend the enormity of the fighting at Paestum, in the first week, from September 9-15, the 151st expended more shells on the enemy than were fired during its entire action in the Tunisian campaign.

For the field artillery, the fighting in Italy could not have been more different from the Tunisian campaign. There was only occasional ground that was open and flat in Italy. The war here would be in large part a hill and mountain war where artillery observation positions were many and excellent for both sides. So tactics were changed. Artillery pieces were positioned at night and carefully camouflaged and protected from daytime counterbattery fire. The few good battery-siting positions found each day forced batteries to clump close together — never a good idea — and even fire over one another.

By October 7, Allied armies — American, British, French and later a Brazilian unit — had rolled back the German defenses to the Volturno River, 20 miles north of Naples. Naples had been taken on October 2. The front at this time, across the peninsula of Italy, was about 80 miles long. The American Fifth Army was assigned to the sector from Benevento southwest to the Tyrrhenian coast, the left side, and the British Eighth Army was assigned the sector from Benevento northeastward to the Adriatic Sea, the right side. The American components included the 3rd, 45th and 34th Divisions, and the attached British X Corps of three divisions. Against these Allied Divisions, the German Commander, Field Marshal Albert Kesselring, commanded six divisions, four armored Panzer and two infantry. The Allies had an advantage in total numbers of troops, 100,000 to 60,000, but the Germans had an advantage in hardened and almost invulnerable mountain defensive positions.

A 155-mm gun (Long Tom) at the Anzio Beachhead in Italy, is fired against a German position. (Photo courtesy of the National Archives)

For the next two months it was a slugfest northward up the peninsula, the Germans grudgingly yielded ground only yard by yard and exacted a terrible toll on Allied attackers. By early November, the Allies had reached the Gustav Line, an enemy winter defensive line across the Appennine Mountains, anchored by Cassino on the left and Ortona on the right at the Adriatic Sea. Cassino was attacked in January 1944, but did not fall until the following May.

On the morning of March 15, 1944, Allied aircraft dropped almost 1,000 tons of bombs on German defenses in the Cassino area, while field artillery and tank destroyers shelled the target after each wave of aircraft. This combined effort reduced to rubble, the town and monastery atop a nearby hill, which the Germans were using for observation. Seven hundred forty-six guns and howitzers then delivered 2,500 tons of high explosives in front of assaulting infantry and an additional 1,500 tons on selected targets. Allied field artillery fired almost 200,000 rounds in eight hours. The Germans found protection in bunkers, caves and tunnels and returned to their defensive positions when the shelling had stopped. The American II Corps and British artillery launched a smoke program to hamper German counterbattery work and to cover moving 155-mm guns, 240-mm howitzers, 8-inch howitzers and other field pieces to within 1,500 yards of the enemy. The Allies now had more than 2,000 guns, including tank destroyers, on a front of approximately 25 miles from the mouth of the Garigliano River to the mountains beyond Cassino. On May 11, 1944, Allied corps and division artillery hit all known and suspected enemy batteries for one hour. Corps artillery and tank destroyers shelled road junctions, command posts, lines of communication, bridges and enemy installations and reinforced division artillery as required. After a week of these intensive Allied bombardments, and with aggressive infantry charges, the Germans finally retreated from Cassino. It took five Allied divisions of infantry and artillery to do the job.

In the meantime, the Allies landed at Anzio on January 22, about 35 miles south of Rome, but German strength of some five divisions and more than 40 artillery batteries kept the beachhead confined to an area no more than ten miles deep for four months before the Allies could break out. The enemy attacked the beachhead again and again, both flanks and the center, but combined artillery and Allied air attacks prevented the toehold from destruction. Massed fires did the job again. The breakout started on May 23, 1944, with an artillery barrage beginning at 0600. For the next ten days, some of the bitterest infantry fighting and most intense artillery bombardments in the entire Italian campaign took place on the road to Rome. On June 4, 1944, the Allies entered the Eternal City. The Germans chose not to defend Rome and evacuated their divisions northward. The enemy was now on the run, and the Allies pursued relentlessly north to the Arno River, which was reached late in July. Here, the American Fifth Army was forced to stop and rest. It had been fighting nonstop since the Anzio/Paestum landings and was simply exhausted.

During the Allied advance northwestward up the boot of Italy, the Germans had time to build three major lines of defense in northern Italy. The first was the Arno Line in the mountains rising from the plains just north of the Arno River then swinging south of Florence and continuing on to the Adriatic Sea. The second was the Gothic Line, beginning about 15 miles north of Pisa in the west, again running eastward through the Futa and Gioga passes in the central mountains, then to Pesaro on the Adriatic. The third, the Appennine Line, ran along the peaks of the high mountains overlooking the fertile Po River valley. This line was about 50 miles north of the Arno River. All these defenses had to be breached to reach the Po Valley. It was hoped this feat could be accomplished before the harsh alpine winter set in, but it could not be done. Allied troops and artillery breached the first two defense lines in the fall of 1944 and reached the Appennine Line in January 1945, where the advance bogged down into patrol actions for the remainder of a bitter winter. The assault on the Appennine defenses began on April 15, 1945, after the spring thaws, preceded by two days of incredibly fierce Allied artillery and air bombardments.

The drive was no walk-over; the Germans realized that once this line had been breached and Allied troops were in the Po Valley, it would be only a matter of weeks before German armies would be cut to pieces by our superior artillery, tank and air forces. Resistance was often fanatical. The Germans made stands on the ridges, on hill tops, in farm houses, in churches and in cemeteries. In the ensuing ten days, a type of chaotic warfare took place on the broad plains of the Po River. The speed of this thrust, north and west, resulted in cutting off large pockets of German divisions. They were everywhere — to Allied flanks, rear and front — resulting in many small but often fierce engagements. The Allies struck north to the Brenner Pass at the Austrian border, and west as far as their supplies and ammunition would take them. The Italian Partisans had liberated Milan and, in the process, had captured and executed Mussolini, who had been trying to escape to Switzerland. German troops poured in by the thousands from all directions to surrender to Allied forces. This was a spectacle seen only once before in the closing days of the Tunisian campaign, which now seemed so long ago. All German forces in Italy surrendered on May 1, 1945. From Dastrup (p. 216):

The field artillery received praise for its action in Italy. "The mass of available artillery of all calibers, skillfully controlled and accurately directed, so effectively neutralized the enemy countermetal [artillery] that the breakthrough of the line of the Garigliano and the break out from the Beachhead were assured," Headquarters, Mediterranean Theater of Operations, U.S. Army, wrote in early 1945. Resembling World War I combat all over again, counterbattery fire from corps artillery silenced enemy guns, destroyed lines of communication and reinforced division artillery while the latter knocked out all obstacles in the path of the advancing infantry. Seeking to destroy determined enemy resistance in prepared defensive positions, the Army massed fire in unprecedented proportions in Italy.....

Besides reaffirming the wisdom of concentrating fire, the Italian Campaign strengthened the importance of the fire direction center and air observation. The center allowed commanders to disperse batteries in the face of difficult terrain and destructive German counterbattery fire, but mass fire quickly and accurately. In a memorandum in March 1945, [Brigadier] General [David G.] Barr wrote that the fire direction center allowed massing of fires up to seven battalions after the adjustment of observed fire by one battalion or by one forward observer to become routine. The center permitted the field artillery to fire devastating TOT [time on target] missions to surprise the enemy and prevent it from escaping into bunkers or other forms of protection.

France and Germany

Two days after the fall of Rome in Italy, the Allies mounted the largest amphibious attack operation in history, the invasion of Europe at Normandy, France, from bases in England. The date was June 6, 1944. The assault on Europe was to begin from a beach, code-named Utah, on the Channel coast of the Cotentin Peninsula, eastward along the Normandy coast, to include four more landing sites, code-named Omaha, Gold, Juno and Sword Beaches. The easternmost, Sword Beach, was at the mouth of the Orne River, north of Caen, while the westernmost, Utah Beach, was about six miles north of Carentan. From Utah to Sword, the assault line was more than 40 miles long. Bradley's U.S. First Army was the American assaulting unit, with the 4th Division in VII Corps, under General Lawton Collins, hitting Utah Beach and the 1st Division in V Corps, under General Leonard Gerow, hitting adjacent Omaha Beach. Men of the U.S. 82nd and 101st Airborne Divisions were parachuted behind Utah Beach during the early morning hours of darkness, just before the landings, to disrupt enemy communications and transportation. With the 82nd Airborne was the 320th Glider Field Artillery Battalion that landed by parachute and glider in the vicinity of St. Mere Eglise, inland and west of Utah Beach. Only two of their 105-mm howitzers made it down safely, but by June 8, eight were in action. The 320th's next landing would be near Groesbeck, Holland, in September 1944, in support of the 82nd Airborne's 325th Glider Infantry. Eight of 12 105-mm howitzers made the Groesbeck landings safely.

The British Second Army would assault the remaining three beaches, with the British 50th Division assigned Gold, the Canadian 3rd Division, Juno, and the British 3rd Division, easternmost Sword. The British 6th Airborne Division would be dropped behind Sword Beach with the objective, as at Utah, of disrupting enemy communications. In the immediate area, opposing the Allied assault, were seven German divisions that, in a matter of a few days, would grow to 22. At least seven of this latter number were Panzer (armored) divisions. Widely dispersed in France and the low countries at this time, were 50 German infantry divisions and ten armored divisions.

D-Day was originally scheduled for dawn on June 5, 1944. The worsening weather dictated otherwise. An approaching atmospheric depression whipped up the Channel seas on the 5th, the rains came, and the cloud base lowered

B Battery, 320th Glider Field Artillery Battalion, prior to D-Day, June 5, 1944. (Photo courtesy of the 320th Airborne Field Artillery Battalion Association)

T-5 Ed Asbury (center) of the 320th Glider Field Artillery Battalion with two other soldiers preparing for the Normandy invasion. (Photo courtesy of the 320th Airborne Field Artillery Battalion Association)

to 500 feet, making air support for the landings tentative at best and impossible for the medium-altitude bombers. General Eisenhower postponed the assault for 24 hours; D-Day was now June 6. Vehicles and guns were water-proofed, anti-gas clothing was issued and the armada started across the Channel in the late evening of June 5 for the early morning landings. The invasion flotilla that approached the coast of Normandy in the early morning hours of June 6, was of gigantic proportions. Some 1,700 warships, 2,700 support ships and 2,500 landing craft bore down on their assigned beaches. Fog and mist shrouded the armada during its approach, and the enemy was caught, initially, by surprise. Then the airborne troops began to land in darkness, enemy antiaircraft guns opened on the troop-carrying planes and the fighting ashore began. The big naval guns offshore now began pounding enemy defenses at the beaches and fighter-bombers streaked in at low level to add to the destruction. Eleven thousand Allied planes would be in the air over northern France on this day.

The first troops to land were the men of the 4th Division, on Utah Beach, at 0630 hours. The 1st and 2nd Battalions of the 8th Infantry Regiment were the first ashore. Elements of the 4th's Military Police Platoon and elements of the 4th's Medical Battalion went in with the first assault waves. They were needed. The 8th Infantry's 3rd Battalion and the 3rd Battalion, 22nd Infantry Regiment, each with a platoon of engineers, were on the beach by 0745 hours. The two engineer platoons opened a road off the beach for their two infantry battalions. The rest of the Engineer Battalion landed later on D-Day and began its work of removing mines, filling shell craters, constructing short-span bridges and assaulting hardened enemy fortifications. First elements of the 29th Field Artillery Battalion, Lieutenant Colonel Joel F. Thomason, Commanding, were ashore by 0915 hours. The 20th Field Artillery Battalion, Lieutenant Colonel Fred W.

Urick, Commanding, was ashore on the night of the 6th. The 1st and 2nd Battalions, 22nd Infantry, followed with elements of the 44th Field Artillery Battalion, Lieutenant Colonel William A. Watson, Commanding, and the 12th Infantry Regiment landed later in the morning along with the 42nd Field Artillery Battalion, Lieutenant Colonel Thomas I. Edgar, Commanding. The 44th fired the first artillery rounds from Utah Beach in support of the 22nd Infantry's attack inland. Leading elements of the 90th Division landed later in the day.

The 1st Division hit Omaha Beach and had rough going. They were immediately pinned on the beach. Every artillery piece, save one of the 111th Field Artillery Battalion, was lost in the assault on the beach. The 7th Field Artillery Battalion lost six of its pieces in this landing. All of these artillerymen became infantry and many died on the beach. With the 1st Division was the 5th Field Artillery Battalion, whose guns were set up and firing by the next morning; it would fight across France and into Germany to the end of the war. The 5th had a long history, tracing its lineage from the Revolution through every war in the country's history. (In 1964, the 5th Battalion became the 1st Battalion, 5th Field Artillery Regiment, and served in Vietnam [towed 105-mm howitzers] and the Persian Gulf War [self-propelled 155-mm howitzers].)

Among the first artillery that went in at Utah and Omaha Beaches were the towed M2 105-mm howitzers, and most of them made it ashore safely. Four days later there would be more than 600 towed and self-propelled pieces on the beaches. The infantry fought landward and ran into the infamous bocage or hedgerow country of Normandy, four-foot earthen dikes surrounding orchards and fields and separated by unmapped dirt lanes and paths. This was perfect defensive country and a nightmare for an attacking infantry that could not see the hidden enemy. It was no picnic for artillerymen, either. The top of the earthen dikes provided temporary good observation at times, but it was a dangerous place for an observer to be. Air observation worked better in the broken country.

One of the first goals was to break out of the bocage, and that meant the town of St. Lô had to be taken first. The terrain was open south of St. Lô, and it was good tank country. The American VII Corps, with three divisions, battled north to take Cherbourg on June 27, but for the 4th Division, it had been costly. In the three weeks following the landings, the division had lost 844 killed and 4,687 wounded. St. Lô fell on July 17, to the V, VII and XIX Corps and the American advance halted briefly to reorganize. During this brutal fighting against German infantry and armored Panzer divisions, the field artillery had to prove itself again and again. Corps artillery shelled the enemy behind the front lines and time and again prevented the Germans from calling up reserves. Division artillery stopped innumerable enemy tank attacks and laid down rolling barrages in front of its advancing infantry. In some battles, massed fires of up to 20 artillery battalions were needed to take out hardened enemy positions. Breakout from the bocage was on July 25, with 1,000 artillery pieces in support of the infantry and tank attacks. More than 100,000 artillery rounds broke the back of the German defenses and silenced enemy batteries. Massed fires were directed at map positions known to be crucial to the enemy; air observation was critical and effective. Central fire control, with observers attached to infantry and tank units, worked superbly; massed fires could be executed with a few minutes notice.

There then began a rout rarely seen in modern warfare. Two American armies (First and Third), the British Second Army and the Canadian First Army began a relentless pursuit of the retreating enemy eastward toward Germany. If the Germans invented "blitzkrieg" it was redefined for them now. Brest fell to Patton's tankers on August 7, 1944; Le Mans

Members of the 41st Field Artillery, 3rd Infantry, fire on Aix-En-Provence, outside the city of Pallette, France, on August 20, 1944. (Photo courtesy of G. William Mohr)

Battery A, 34th Field Artillery Battalion, 9th Infantry Division, pulling a 155-mm howitzer out of position to move on to a new position in Germany. (Photo courtesy of Robert C. Baldridge)

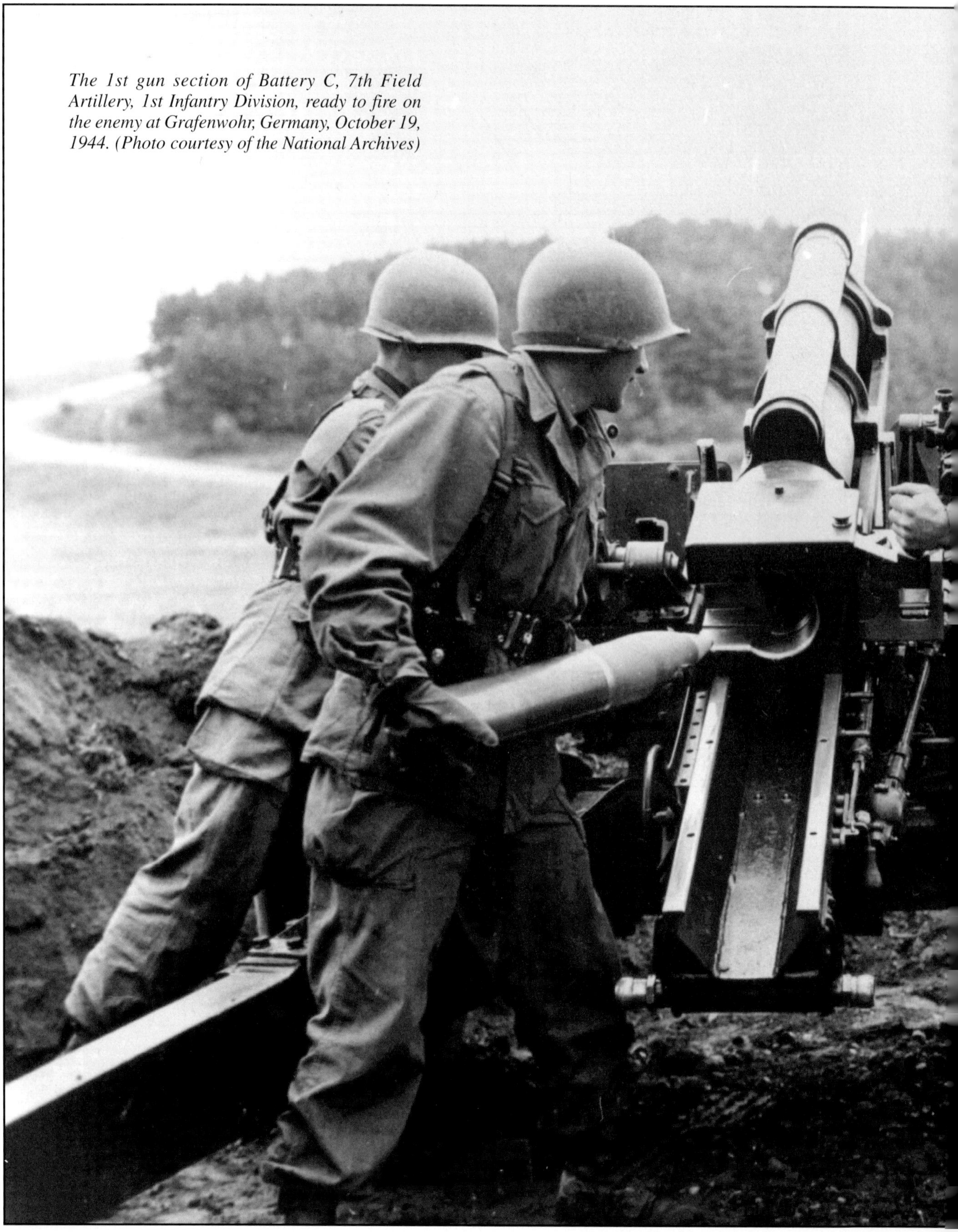

The 1st gun section of Battery C, 7th Field Artillery, 1st Infantry Division, ready to fire on the enemy at Grafenwohr, Germany, October 19, 1944. (Photo courtesy of the National Archives)

on August 8; Orleans and Chartres on the 16th; Mantes on the Seine River on the 19th and the biggest prize of all, Paris, on August 25. Field artillery had little chance to support this fast moving tank/infantry drive, for the enemy, except in isolated instances, chose not to stand and fight. But this headlong drive slowed and halted in September as the Allies swept through Belgium and Luxembourg to the Dutch border. Supply lines had almost overnight become very long and unmanageable. German defenses also stiffened as the Allies approached the German border in the center and right sectors. Heavy artillery pieces were then moved up to the front as the fighting quieted temporarily. Patton's Third Army tankers and infantry captured Nancy on September 15, but supply lines were long and slow; ammunition and fuel were now in short supply.

On August 15, three American divisions and four French divisions assaulted the southern French city of St. Tropez on the Mediterranean coast. The American divisions were the 3rd, 36th and 45th, part of General Alexander M. Patch's Seventh Army. With the 3rd Division were the 9th, 10th, 39th and 41st Field Artillery Battalions. These divisions fought northward, generally on a path between the Rhone River and the Italian border with France, then into the Vosges Mountains, eventually driving into Germany south of Strasbourg seven months later, on March 15, 1945. In one instance, in the first weeks after the landing in France, 3rd Division's artillery caught a 1,000-truck, 8-mile-long German convoy retreating north along the Rhone and almost completely destroyed it. The Seventh Army would eventually link up with Patton's Third Army for the final drive into Germany.

Early in November, XX Corps, on the right, jumped off at Metz toward the German border, about 30 miles east. Supported by 700 artillery pieces that smoked enemy observation positions and shelled all known enemy strongpoints, pillboxes, forts and enemy batteries in a huge

Gunners of Battery A, 19th Field Artillery, 5th Infantry Division, near Haller, Luxembourg, are in position with shells ready for rapid firing of a 105-mm howitzer on January 12, 1945. (Photo courtesy of the National Archives)

barrage, Patton's infantry and tanks moved out to take Metz, and by December 15, had crossed into Germany in the Saar region. It was a much slowed pace, and now the force of Allied artillery could be brought to bear to prepare the way for the American tank and infantry thrust. One hundred thirty thousand rounds from 105-mm and 155-mm howitzers battered the enemy in Metz and the Germans were forced to withdraw. The American First Army in the north, assaulted Aachen, Germany, on October 11. For ten days, tanks and artillery, 155-mm howitzers and 155-mm guns, battered the city in support of vicious house-to-house street fighting. Some 75 batteries at Corps and Division level (about 300 pieces) reduced the city to rubble, and the enemy was driven to the banks of the Roer River. By December 15, 1944, all Allied armies, except the British in the Netherlands, were up against the German border and across it in a few places. Now the winter turned very cold and deep snows buried the battle zone.

In a last gasp of offensive bravado, the Germans on December 16, 1944, hit a weak American front in the Ardennes Forest with three Panzer Armies — 38 divisions in all. For ten days, this juggernaut rolled into Luxembourg and France to a depth of about 50 miles to near Dinant and Ciney, almost to the Meuse river, and encircled (but were unable to take) Bastogne. This was the so-called "Battle of the Bulge." In the initial attack, American infantry, new at the front and untested, could not stop the German thrust. Artillery forward observation posts were overrun and American batteries were forced to withdraw. But at Monshau on the far right German flank, 15 battalions of 105-mm, 155-mm, 240-mm and 8-inch howitzers, 4.5-inch and 155-mm guns stopped the drive in that sector. And all along the German flanks, artillery from some 12 Allied infantry divisions and at least six armored divisions from five corps gradually funneled the German armies into a narrower and narrower point from the original 50-mile-wide attack front. Artillery had a chance to try out the new variable time (VT or Pozit) fuzes on their shells, which exploded at predetermined distances from a target using reflected radio signals. They were deadly and were in common use for the remainder of the war in Europe.

Finally, short of ammunition and fuel and decimated by massive Allied artillery barrages often with direct fire, including 4.5-inch rocket fire at 100 rockets per minute, the German attack stalled. Late in December, the First and Third Armies attacked the German flanks and the enemy turned eastward in a desperate rush to avoid being trapped in a pocket of its own making. The 11th and 2nd Armored Divisions met at Houfallize in January to close off the "Bulge." Patton's Third Army was directed north to help relieve Bastogne. All during the encirclement of Bastogne in December 1944 and January 1945, in freezing temperatures and on snow-covered ground, the 420th Field Artillery Battalion and the 10th Armored Division pounded the attacking Germans with their 105s

Artillerymen of the U.S. Third Army take time out to enjoy coffee and listen to the harmonica, in Bedersdorf, Germany, March 1, 1945. (Photo courtesy of the National Archives)

on the west side of the town, stopping attack after attack until the Germans were forced to withdraw with the main body.

Hitler's tankers and infantry escaped back into Germany before the Bulge pocket could be sealed off. But there were inexplicable atrocities: at Malmedy 82 captured men of Battery B, 285th Field Artillery Observation Battalion, were simply machinegunned to death in the snow in an act of killing without reason. By late January, all lost ground had been recovered by the Allies and they were again on the German border. One hundred ten thousand of the enemy were captured, and they suffered an additional 100,000 casualties. The attack had gained them nothing.

War's End in Europe

The Allies picked up the tempo of the fighting early in March 1945, and the British and Canadians from the north, the Americans and French from the west and the Russians from the east, converged on Germany. The German defenses stiffened on his own soil but quickly crumbled in the path of the massive Allied onslaught. Holland was liberated by the British and Canadians, and virtually all of central and southern Germany was overrun by four U.S. armies. Adolf Hitler shot himself on April 30, rather than be captured. Berlin fell to the Russians two days later and Germany surrendered, unconditionally, on May 7 at Reims, France. The following day was celebrated as Victory in Europe (VE) day.

Firepower and maneuver were judged to be the fundamental elements of combat in Europe. Operations were successful when ammunition was plentiful and when massed artillery fires were employed. To prevent the slowing of operations, the field artillery required an unrestricted supply of ammunition. On rationing, the American Sixth Army Group commented that a larger supply of rounds for the field artillery would have reduced friendly casualties, would have killed more Germans and would have ended the fighting sooner. Although the field artillery had adequate supplies of ammunition, field artillerymen wanted even more ammunition since nothing substituted for massed fires and the lavish use of rounds.

The fire direction center, organic air and ground observers, and self-propelled and towed artillery brought a new dimension to the field artillery. Motorized field pieces gave the field artillery mobility to stay abreast of fast-moving infantry and armor columns. The fire direction center permitted massing fires, shifting fires rapidly and accurately around the battlefield, keeping field artillery battalions dispersed to protect them from counterbattery fire, exploiting the available ammunition and supporting the other combat arms.

The War in the Pacific

Following the attack on Pearl Harbor in December 1941, the Japanese overran one Pacific nation after another in rapid order. Imperial Japan had already attacked and occupied Manchuria on the Asian mainland in the early 1930s and invaded China proper in 1937, controlling much of eastern China before the end of that decade. French Indochina (Vietnam) fell to the Imperial armies in 1941, followed during the first half of 1942 by Burma, the Philippines, the Netherlands East Indies, Borneo, Celebes, much of New Guinea and as far eastward as the Solomon, Gilbert and Marshall Islands. The Japanese landed in the northern Philippines on December 10, 1941, and at Lingayen Gulf, Luzon Island, on December 22.

The American artillery unit defending Luzon Island was the Provisional Field Artillery Brigade, formed from the Headquarters of the 24th Field Artillery Regiment, Colonel Louis R. Dougherty, Commanding. The brigade consisted of nine battalions of artillery, most armed with 75-mm guns; three battalions of the nine were self-propelled and five battalions were equipped with towed pieces. One other was a much modified 155-mm GPF battalion. By early January 1942, these battalions and Army infantry had been bottled up in the Bataan Peninsula south of Manila and began a delaying action that has few equals in military history. About 110,000 Americans and Filipinos were on Luzon at the time of the Japanese landings; by December 31, there were 15,000 Americans and about 65,000 Filipinos left. From about January 6, these brave units held off attack after attack of the Japanese forces on the peninsula, destroyed several Japanese amphibious landings until, with relief impossible and with little ammunition, fuel or food left, they destroyed their own weapons, pushed them into Manila Bay, and on April 9, 12,000 Americans and 64,000 Filipinos surrendered. The Bataan Death March of 65 miles followed, during which thousands died.

The tiny American garrison at Wake Island, 750 miles northwest of the Marshalls, surrendered on December 23, 1941, and the occupation of three barren islands, at the end of Alaska's Aleutian Islands chain, marked the pinnacle of Japanese expansion into the Pacific. By mid-1942, virtually all Pacific land and sea areas west of the International Date Line and north of 10 degrees south latitude were firmly in Japanese hands, fortified and strongly defended.

Solomon, Gilbert, Marshall Islands and Guadalcanal

After several naval battles in which American forces, although outnumbered, were victorious, the island-by-island rollback of Japan began by American Marines who attacked Guadalcanal in the Solomon Islands on August 7, 1942. Artillerymen in the Pacific used the same types of weapons as were being used in the European war, but this was an island war in the Pacific, fronts were restricted, most combat was in jungles and at close range, and so artillery tactics were required to be quite different. At ranges often less than 2,000 yards, artillery on Guadalcanal used direct fire to dislodge the Japanese from strongpoints to support Marine drives through the island. High-angle, short-range fire was necessary in jungle warfare, and the absence of any discernible front line required batteries to protect themselves from infiltrators in a 360 degree circle. Few of these tactics had been taught them, but they were used on every island invaded; it was on-the-job learning. The final American drive on Guadalcanal started in December. The Marines and Army used every piece of artillery they had: 75-mm pack (mountain) howitzers, 105-mm and 155-mm howitzers and 155-mm guns — some 80 pieces in all. The Japanese finally decided the island wasn't worth the devastating casualties they were taking and evacuated Guadalcanal on February 8, 1943.

Island after island now began falling to the Army and Marines as the troops learned jungle warfare. First landings in south New Georgia, about 150 miles northwest of Guadalcanal, by units of the 43rd U.S. Infantry Division (Major General John H. Hester) and the Marine Corp's 4th Raider Battalion (Colonel Harry B. Liversedge) took place on June 21. Two weeks later, part of the 1st Marine Raider Regiment and reinforcing elements of two more infantry divisions (the 37th and 25th) went ashore in New Georgia to overcome stubborn enemy resistance. Nearly 30,000 Americans were fighting in this campaign (including those on the surrounding islands of Vangunu, Rendova and Arundel) before New Georgia and these islands were finally secured in September. Eastern New Guinea, near Lae, was invaded by the 503rd U.S. Parachute Regiment and the Australian 7th Division on September 4, and these and other units pushed toward this major north-coast port.

Bougainville and Tarawa

On November 1, 1943, the 3rd Marine Division and the Army's 37th Infantry Division assaulted Bougainville in the Solomons and began three weeks of vicious fighting, and on November 20, the 2nd Marine Division hit Tarawa Atoll in the Gilberts. At Tarawa, the low tide grounded many of the landing boats and the Marines were forced to wade ashore under murderous enemy fire. Five sections of the 10th Marines' artillery landed after the initial assault force, but three foundered in the shallow water, and the artillerymen carried their disassembled pack howitzers to the beach on their backs. Two artillery battalions eventually made it ashore. It took three days to subdue the atoll's defenders, and on one of these nights, November 22, the 10th's 1st Battalion stopped a night banzai charge with 1,500 rounds, killing 325 of the fanatic attackers. The Marines suffered 3,500 casualties on Tarawa.

New Britain was invaded in December 1943; unoccupied Majuro fell on January 30, 1944, and Kwajalein, in the Marshall Islands, fell on February 4 to the 4th Marine Division and the Army's 7th Infantry Division. Eniwetok, also in the Marshalls, was cleared on February 21 (the U.S. suffered 3,000 casualties on two atolls). Troops landed in the Admiralty Islands north of Rabaul this same month. Through the spring and early summer of 1944, MacArthur's army took one Japanese base after another along the north New Guinea coast, reaching the far western end of the huge island by August 1944. The gigantic enemy bases at Rabaul and Truk were neutralized by land- and carrier-based air bombardment and bypassed.

Mariana and Palau Islands

The U.S. Navy now made a 1,200 mile jump to the northwest to the Mariana Islands, about 1,500 miles south of Tokyo. These islands were to become long-range bomber bases, bombers that would eventually decimate the Japanese home islands. The next island for the 10th Marines was Saipan in the Marianas. This island was assaulted on June 15, 1944, by the 2nd and 4th Marine Divisions with 20,000 men and by the Army's 27th Infantry Division, which landed on the 17th. The Japanese had 30,000 men defending Saipan with about 60 artillery pieces. The 10th Marines set up their pieces on the beach with little cover. On June 23, the 10th Marines artillery command post was shelled by an enemy battery, killing three men. The Regiment's 1st Battalion assumed fire control for all battalions. As the fighting began to wind down in July, the Japanese defenders became more desperate. At one time, the fighting was so close that the 10th's 3rd Battalion set its fuses to detonate 50 yards from their batteries. The 25 days of fighting on Saipan were bloody and bitter; the Japanese made the Americans pay for every yard gained. Fourteen thousand Americans were killed and wounded, and almost the entire enemy garrison was killed. Saipan was secured by July 9.

Tinian, another of the Marianas, was next to fall to the 2nd and 4th Marines who assaulted the island on July 24. The Japanese garrison on Tinian numbered about 9,000 men. Tinian was shelled by the 10th Marines and other artillery units from the south end of nearby Saipan with 156 heavy guns, many manned by Marine units, and the 10th also landed 75-mm pack howitzers on the first day. Again, point-blank direct fires were often required of both Army and Marine artillery to stop fanatic enemy suicide charges. Tinian was secured on August 1, and most of the enemy garrison were killed in the eight-day battle.

Guam, about 150 miles south of Saipan, was assaulted by the 3rd Marine Division, the 1st Marine Brigade and part of the Army's 77th Infantry Division on July 21, 1944. Fifteen thousand Japanese garrisoned the island. As on the other Mariana islands, the enemy put up a fanatic resistance. Eight battalions of artillery and Marines and Army infantry were required to

Camouflaged 105-mm howitzers of 245th Field Artillery, Americal Division, blast the Japanese positions on Negros Island in the Philippines. (Photo courtesy of the National Archives)

take the Orote peninsula, which fell after six days of fighting. One thousand four hundred Marines and soldiers died on Guam; the defenders lost at least 10,000 killed and possibly more. Guam was declared secured on August 10.

The next island group was the Palaus, 550 miles due east of Mindanao, Philippine Islands, which included Peleliu, Babelthuap and Angaur. The 1st Marine Division and the 81st Infantry Division were the assaulting forces, and they ran up against some of the most fanatic resistance of the Pacific War to that date. Clearing the Palau Islands took the Marines and Army more than a month, from September 15, to near the end of October 1944. Artillerymen had to improvise, again, in the Palaus. Some regimental batteries fired into other unit's zones to take advantage of the guns' range and firepower. Marine howitzers lined up side by side on their restricted beachheads and bombarded at close range enemy pillboxes, caves, ammunition dumps, dugouts and the usual bridges and buildings. The Palaus were bloody; Americans suffered more than 10,000 casualties. Most of the 30,000 Palau defenders were killed.

Philippine Islands and Leyte

On October 20, 1944, the U.S. Sixth Army, commanded by Lieutenant General Walter Krueger, invaded Leyte Island, Philippine Islands, in a bitterly contested campaign that continued into July 1945, before all the numerous islands were cleared of the enemy. Americans killed in action in the Philippine campaign were 3,504; Japanese casualties, however, were known to be at least 15 times this number. Artillery tactics underwent another change in the Philippines. Long pre-assault bombardments, it turned out, achieved little in the way of decimating enemy defensive troops. The enemy simply went under cover until the bombardment stopped and then came out to face the assaulting U.S. infantry. To foil this Japanese tactic, artillery fired short preparations, ceased for some time to allow the enemy to come out from cover, and then sent in another barrage, catching them in the open. It worked time after time. The Japanese seldom massed fire on American batteries, so American infantry was always well-supported by their artillery. But the Japanese also fought for every yard of ground, preferring to die rather than suffer the humiliation of capture. American artillery became more and more important as a tool to dislodge fanatic defenders. Some 90-mm antiaircraft artillery (AAA), serving as field artillery, was used in direct fire against hardened enemy bunkers. Hard fighting continued through November and December on Leyte Island and finally on January 4, 1945, the Japanese were driven into the mountains in isolated groups, and their effectiveness as a fighting force ended.

The field artillery played a conspicuous role on Leyte. Aided by radio-equipped ground and organic air observers and fire direction centers, artillerymen were able to mass fires and follow closely the infantry and armor despite the constant mud and rain. The terrain often forced artillerymen to emplace their batteries along small fronts and fire over each other from ranges as close as five hundred yards from their targets. Even though the field artillery had difficulty destroying enemy defenses, and at times faced ammunition shortages because of the problems of transporting ammunition through mountainous terrain, massed artillery fire disrupted enemy communications and observation, prevented the construction of additional defensive positions in depth and disrupted the movement of enemy reserves.

Luzon

The Sixth Army then struck Luzon Island at Lingayen Gulf on January 9, 1945. Clark Field was taken on January 26. American troops landed at several places on Luzon and on Samar and Palawan Islands during February 1945 (Mindoro Island had been assaulted in early January). Manila and Corregidor fell on March 3, Negros in June and Mindanao in July. On many of these large islands, the enemy dug in and defended caves and other strongpoints that artillery could not reduce. As a result, most operations became infantry flame-throwing and small-arms battles at close range during the spring and summer offensives in the Philippines. The exception was Manila itself, where the enemy defended every building in almost room-by-room combat. To take the city, artillery — 155-mm and 8-inch guns — would half-ring a building and reduce it to rubble at point-blank range before the infantry attacked with flame throwers, grenades and small arms. Even then, the few survivors would not surrender and chose to die.

Iwo Jima

Between February 19 and March 26, 1945, little Iwo Jima in the Volcano Islands, 600 miles south of Tokyo and halfway between the bomber bases in the Marianas and Japan, became a killing ground for the men of three Marine Divisions, Navy and Army; 6,821 Americans, mostly Marines, died on Iwo Jima. Almost all 20,000 of the defending Japanese garrison chose to die. No longer would enemy fighter planes rise from Iwo to attack Army Air Force B-29s on their way to bomb Japan, and thousands of crippled bombers and their crews would find a safe haven on Iwo Jima's airstrips. In this battle, artillery, as on some other fanatically defended islands, could not dislodge the underground enemy, for the entire defending force fought from caves and tunnels throughout Iwo Jima. This was an infantry job, with flame throwers and satchel charges the preferred weapons. Thunderous offshore naval bombardments, Marine artillery fires and Marine and Naval air bombardment made a moonscape out of little Iwo Jima, but had virtually no effect on the dug-in enemy except to keep him underground. Major Marine artillery units at Iwo Jima were the 12th Artillery Regiment (3rd Division), 14th Artillery regiment (4th Division), 13th Artillery Regiment (5th Division) and the 1st Provisional Field Artillery Group (2nd and 4th 155-mm Howitzer Battalions) of the V Amphibious Corps.

Warrant Officer, George J. Green led a team of Forward Observers for artillery on Iwo Jima. He was with F Battery, 2nd Battalion, 12th Marine Regiment (Artillery), and was with the 3rd Division from February 21 when he came ashore, until March 17, when his unit was shipped out. Some vivid recollections of his 25 days on the island are worth reading here:

February 24. D+5. Good night's sleep — only woke up once during the air raid. The Company area was hit by mortars last night but none of my Forward Observer team was hit — several casualties in K Company.

It was still dark but dawn was beginning to break when we crawled into the Jap pillbox that Lieutenant Colonel W. DuPlantis, Commanding Officer, 3rd Battalion, was using as his Battalion Command Post. I know we reviewed the situation, and what we were supposed to do — but the only specific thing I remember is his saying — "We have got to get the airfield today."

We started the attack this morning with about 220 men, and before we got back in reserve, K Company had around 90 men left. We are to pass through the lines of the 1st Battalion, continue the attack and get airfield Number 2. We move into position and take off behind

A Marine sitting on a gun on the beach of Iwo Jima. (Photo courtesy of the National Archives)

a rolling barrage. We now have moved up to the runway, and I keep the team spread out in the holes. We are doing pretty good on communications, as we are in touch with Lieutenant Pottinger at Battalion. He wants to know why we are not doing any firing — I told him I could not see anything to shoot at, and troops are moving up fast. He said pick out some prominent landmark and fire away. I set up a fire mission on the high ground where the runways cross and fire one round of smoke. Just as it hits, Captain Heinze yells they are going across the runway and heading for the high ground. I call off the fire mission having fired only one shell. The next thing is while looking back to our rear, I see all these 4.5-inch rockets coming down at us. They are easy to see as there are so many of them, and they are making such a loud racket. We are now in a good, deep shellhole and sit helpless watching those rockets creeping up on us and exploding on our reserve and machinegun sections. Finally they stop — the last one being about 25 feet behind us. I heard later this barrage by our own rocket trucks just about wiped out the Company machinegun section.

Early in the evening, elements of the 2nd Battalion came up to fill in the gap on our left. It seems that K and I Companies were the only ones to get across the runway. Along with them came Warrant Officer Cormier and his Forward Observer team from D Battery. We never did make contact with the 5th Division on our left that night, and I'm positive they tried to cover this gap with artillery, and some of it was landing on us. We were not having a lot of luck without telephone communications. I fired in some barrages and got set up for the night. Still not much success with radio though.

With the good tidings of food, ammo and water [delivered that night] also came the loss of our telephone communications; since radio was so undependable we decided to get the wire fixed. I sent Corporal McClosky and one man out to see if they could find and repair it by following the wire from the telephone, back. It was really dark now — you couldn't see too far. They took off without hesitation but were back before too long with no luck. They didn't know how far they had gone out. Then Sergeant George Chest and I started out along the wire, and after going what we guessed to be about 100 yards, we found the break. George repaired it and we got back to our position without difficulty. How and why that wire didn't break but that one time that night I'll never know as we were being mortared all night long.

It was a simple matter to locate ourselves because we were on the hill where the two runways of airfield Number 2 crossed [this airfield was under construction]. Being sure of our location, we realized that as long as we fired our artillery, the Japs let up on their mortars that were falling all over us. So, Chest and I got under our ponchos and started firing by using the target squares on the map for targets and kept moving the shells around in front of our lines. Not too long, in scampered a rifleman asking us to move the shells out a little farther as they seemed only to be 50 feet in front of them. Lieutenant Archambault and I decided not to come any closer, but not to change what we had now. I often wondered if that infantryman ever knew we didn't move the firing out farther.

About 3:00 a.m. I got a call from Captain Karch, who was the S-3 of the Battalion of the 4th Division Artillery. I was firing and he complained about my using up all their ammo. And, since all of my firing was unobserved, did I know what I was shooting at? I told him all I knew was that when I fired they didn't, and I knew where I was, there was nothing in front of us but Japs and they, Battalion, could get more ammo. After the discussion, we decided to reduce the number of rounds for effect by Battalion.

On occasion we got heavy artillery exploding in our lines. I was convinced it was coming from our positions in the rear. I reported this to the 14th Marines. I think the 5th Division was trying to fill the gap between the 2nd Battalion on our left and their right flank unit and was not aware that we were as far left as we were.

When daylight finally came, the Japs let up on their mortaring and George Chest and I looked around for a more satisfactory spotting position to try and determine where all this stuff was coming from. The Japs had a rocket of large size that you could see taking off. Evidently it was difficult to control the range of these rockets, as earlier in the battle we could see them arch up and fly over Mount Suribachi. Almost everybody saw them because of the noise they made. In our new position we could see these rockets taking off and reported the compass angle from our position. I could not reach them with my artillery as they were in the 4th Division's zone of action......

I find a nice flat spot behind some rocks at the bottom of a hill and put Cooperman there with the radio. He has radio contact with Battalion and they advise me we are to attack again at 1600 and wants me to fire in the preattack barrage. I find Captain Stevenson and tell him the news; he hadn't received the word from 3rd Battalion yet and doubts if we can be ready by then as we are getting hit pretty hard. I'll be darned if I can see any live Japs. I find an observation spot on top of the hill in front of the radio position and start to register in for the barrage. This required me to spot the round, run down to the radio, give the correction and run back up the hill to see where the shell lands.

We spent the night dug in among the jagged rocks firing unobserved artillery fire at military positions indicated on our maps. Early next morning we were advised our Forward Observer team would be relieved, and by noon we were on our way back to the battery positions.

Okinawa

On April 1, 1945, Lieutenant General Simon B. Buckner's Tenth Army landed 50,000 men on Okinawa, supported by guns and howitzers of the 420th Field Artillery Group (three battalions) on an adjacent island. The landings were unopposed by the Japanese. But the enemy's orders were to fight to the last man, and to the last civilian man, woman and child on the island, for they knew the fall of Okinawa would open the way to an invasion of the home islands and total defeat. The island's rugged terrain made it a defensive paradise for the enemy, as was Iwo Jima. The American force would eventually involve four Army Divisions, the 7th, 27th, 77th and 96th, and four Marine Divisions, the 1st, 2nd, 5th and 6th, totaling almost 550,000 men of which 180,000 were combat troops. It was the most massive amphibious operation in the Pacific War.

The island is 65 miles long and is variable in width from three to ten miles. The initial beachhead was established by two Army Infantry Divisions, the 7th and 96th, and two Marine Divisions, the 1st and 6th. The Americans quickly built up an enormous attack force on a beachhead nine miles wide and three miles deep and, as usual, opposition to the landings was light. The two Army divisions, now joined by the 2nd and 5th Marines Divisions, turned south and had a difficult time of it as they ran up against stiff enemy defenses on the southern ten miles of the island. The other two Marine divisions, the 1st and the 6th, turned east and north, respectively, toward the remaining 55 miles of Okinawa.

On April 2, the two Army Divisions and the two Marine Divisions started the drive inland with three supporting 155-mm howitzer battalions of the 419th Artillery Group. But the Japanese made this a different kind of island war. They concentrated their forces, including artillery, in the south end of the island where they built several strong defensive lines and trained their artillery crews to protect adjacent batteries, in addition to their own. When U.S. troops of the 96th Division attacked, they were met with a devastating barrage of enemy artillery, a new and unexpected tactic, which forced dispersal of the Division's artillery batteries. Now, Corps counterbattery work became important. Forward observers directed fire on single enemy guns and dug-in tanks, and American infantry charged other enemy guns and tanks that artillery failed to knock out. The infantry could not penetrate the first defensive line despite massive bombardments by 27 artillery battalions on April 19. More than 300 pieces were placed about 25 yards apart along the front. But constant attacks by infantry and massive artillery fires finally forced the enemy back by April 24. The attacks continued into May when the American Infantry Divisions and Marines finally outflanked the defenders on both sides and forced a rollback. On April 23, the Army's 77th Division began replacing the exhausted 96th Division.

"Kilo" Battery readying a 105-mm howitzer on Okinawa in 1945. (Photo courtesy of the Headquarters Battery, 11th Marine Division)

L Battery loading round number 75,000 on Okinawa on June 20, 1945. (Photo courtesy of the Headquarters Battery, 11th Marine Division)

The other two Marine divisions, the 1st and the 6th, turned east and north on Okinawa. By April 7, the 6th Marine Division had fought its way north some 20 miles from the beachhead against determined Japanese resistance and began to attack the strongly defended Mobotu Peninsula. After 12 more days of savage fighting in rugged terrain, the Marines cleared the peninsula. By April 13, some units of the 6th Marines had reached the extreme north of Okinawa at Cape Hedo.

The Army and the other two Marine Divisions, the 2nd and 5th, were having a difficult time in the south around the Shuri defense line during the 1st's and 6th's push east and north. Okinawa's capital, Naha, was captured on May 29. On June 4, elements of the 6th Marine Division made an amphibious landing on the

Oroku Peninsula south of Naha, and by the 15th, what remained of the battered 1st Marine Division was integrated into the 2nd Marine Division. General Buckner was killed by an enemy artillery shell on June 18, and the Japanese commander on Okinawa committed suicide on June 22. Okinawan enemy troops then began to surrender en masse, 7,400 of them. The Japanese had suffered at least 110,000 dead, in addition to many civilians who committed mass suicides, having been told the U.S. soldiers and Marines would kill them all anyway. Army and Marine casualties were 40,000, plus another 9,700 Navy casualties, mostly from Japanese suicide plane (kamikaze) attacks.

Japanese artillery tactics on Okinawa were new to the attacking Americans. The enemy, for once, actually employed heavy and sustained concentrations of fire on U.S. positions, forcing the Americans to disperse their batteries. The Japanese, however, did not have enough artillery pieces or ammunition to mass fire effectively. For the Americans, massed artillery bombardments, as in the Philippines, had only a minimal effect on troops hiding in caves, underground or in hardened strongpoints. The enemy had long since learned never to expose itself to American artillery. Naval bombardments silenced enemy rear-area artillery, but could not help the front-line infantry. Over the nearly three months of fighting, however, the Japanese were so badly battered by Army and Marine artillery of every conceivable caliber, and by unending infantry attacks, that they were forced to withdraw into a small pocket against the sea, where those that did not surrender were annihilated. This was the Army's and U.S. Marines' final major battle of the war. The 10th Marines (Artillery) was among the occupation force in Japan, at Nagasaki, for nine months beginning in September 1945, and ending in June 1946. Following occupation duties, the 10th shipped to Camp Lejeune, North Carolina, their permanent home base. They would not be allowed to rest long.

War's End in the Pacific

Okinawa was the final battle in the Pacific. Although American artillery was now receiving heavier pieces for the invasion of the Japanese home islands, an invasion would not be necessary. On August 6 and 9, 1945, B-29s dropped one atomic bomb on each of two Japanese cities, Hiroshima and Nagasaki, obliterating both cities, and the Japanese surrendered. World War II ended with the signing of the surrender document on September 2, 1945, on board the battleship *Missouri* in Tokyo Bay.

Of American artillery in World War II, Dastrup says (p. 236):

Although field artillerymen had to adapt to conditions on the Pacific islands that differed significantly from those where they had been trained, they generally applied standard field artillery tactics. Employing the fire direction center, ground and organic air observers and field artillery group, they massed fire, expended vast quantities of ammunition, taxed the logistical system's ability to supply sufficient quantities of rounds and helped push the Japanese out of their strong defensive positions. Unlike their counterparts in Europe, gun crews in the Pacific used high-angle fire, fought at closer ranges, and faced less aggressive counter-battery fire.

Except for the summer of 1944 when the Allies were chasing a fleeing enemy, artillerymen did not displace their pieces often during a battle. Rather, they moved their weapons forward as weather, terrain and combat permitted, overpowered the enemy with intense massed fires and bombarded their way through highly fortified enemy defenses. American artillerymen overpowered the Germans and Japanese because they could move the fires of large numbers of pieces around the battlefield and concentrate enormous quantities of ammunition on a target.

A Brief Peace 1945–1950

Following World War II, several conferences examined the performance of artillery in Europe and the Pacific, and the overriding conclusions were the need for overwhelming firepower and centralizing of command to control it. To these ends, self-propelled artillery was to completely replace towed pieces, plus add one 155-mm howitzer battalion to the infantry division. Guided missiles were endorsed as the weapon of the future to possibly make obsolete all heavy tube artillery. In 1950, the Army Reorganization Act combined the Coast Artillery and Field Artillery — and their schools — into a single branch. During the first post-war years, the Army experimented with guided missiles, with and without atomic warheads, and developed guidance systems. Out of this came the Corporal and Hermes missiles, quite sophisticated for their time, along with doctrine for employing missiles on the battlefield. By 1950, the Army was ordered to develop tactical nuclear warheads for guided missiles and multiple-rocket launcher capability for the battlefield.

Tube artillery continued to be developed also, if only on paper at first. Prototype 105-mm and 155-mm howitzers and guns were lighter and more powerful than their World War II counterparts, but budget cuts were the bane of the Army's artillery development programs and none of these was fielded. These budget cuts did not curtail reorganization of artillery at the division and corps level, however. That could be done cheaply and in spite of budget cuts. So, through 1949, no new weapons were adopted. There was only experimentation and talk.

Experimentation of dropping heavy artillery pieces by parachute took place at Fort Bragg, North Carolina. The first piece, seen in the photo, barreled into the ground. The second drop was equally disappointing. However, it was the beginning of the concept of heavy drop, now an accomplished task. Photo taken in July of 1948. (Photo courtesy of the 320th Airborne Field Artillery Association)

Korean War 1950–1953

On June 25, 1950, seven North Korean divisions assaulted across latitude 38 degrees North, its border with South Korea, and within three weeks, these Communist troops had engulfed half their far weaker southern neighbor. The United Nations (U.N.) condemned the attack and the war became a U.N. war. U.N. forces were powerless to stop the advance in spite of a heroic stand by the 6th Republic of Korea (R.O.K.) Division and an American task force of 500 men with artillery (part of the 21st Infantry Regiment). By August, the North Korean divisions had advanced to a perimeter around the port of Pusan on South Korea's southeast coast, bottling up U.N. forces in a 40- to 50-mile radius pocket against the Sea of Japan. American troops and artillery that had been rushed to South Korea from Japan were overrun. Reinforcements (mostly American) were flown into the Pusan perimeter, and the North Korean advance was stopped there. The 10th Marines, redesignated the 11th Marines, were rushed across the Pacific to South Korea in time to set up for the breakout from the Pusan perimeter.

Back home, America mobilized again. The Artillery Officer Candidate School was reactivated at Fort Sill, now a course of 23 weeks. General MacArthur was put in charge of all U.N. ground operations and, in typical MacArthur fashion, attacked. On September 15, the U.S. 5th Marines led a beachhead assault at Inchon on the Yellow Sea only three miles from Seoul, South Korea's capital, and far behind enemy lines. The attack was coordinated with a breakout from Pusan, and the enemy, taken by surprise, was swept out of South Korea. Part of the 11th Marines was with MacArthur in these landings. U.N. forces did not stop at the 38th parallel, but continued the drive northward to take the entire southern two-thirds of North Korea, and were only 60 miles from the Yalu River and the Chinese border by the end of October. One U.S. unit actually reached the Yalu on October 26 (17th Regimental Combat Team, 7th Infantry Division). In four months, the aggressor's initial successes had turned to disaster. Then, late in November, 200,000 Chinese Communist "volunteers" stormed across the Yalu and overwhelmed the U.N. forces. In spite of horrendous losses inflicted by U.N. ground troops and air strikes on the attacking Communists, the Chinese pushed U.N. forces south toward the 38th parallel, in some engagements climbing over their own dead troops. The 11th Field Artillery was at the Chosin Reservoir battle at this time, deep in North Korea, supporting the breakout through encircling Chinese Communist lines. By December 15, the front was back to latitude 38 degrees and the Chinese were still attacking.

By January 1, 1951, 400,000 Chinese troops were on the battlefield and had pushed U.N. forces back to Seoul, which was evacuated for the second time. The enemy advance was finally stopped 50 miles south of Seoul, and in February 1951, the U.N. started a slow advance northward to retake the capital. U.N. and Communist forces now began a war of attrition, oscillating back and forth across the 38th parallel, a war that went through the summer of 1951, in offensive after counteroffensive with few decisive victories for either side, although Communist losses were staggering. Cease-fire talks were initiated in July as the Communists needed time to resupply and consolidate their positions. The talks soon broke off and the fighting resumed. U.N. forces attacked northward in September and the Chinese forces slowly gave ground until, unable to stop the U.N. drive, asked for another round of peace talks. On November 23, 1951, a cease fire was declared and all through 1952, while talks continued, a sort of trench warfare in the hills resulted in a stalemate punctuated with mostly small-scale raids by each side.

Action of the 780th Field Artillery (8-inch howitzers) during January 1952 is of interest. The Chinese were observed building a string of some 40 bunkers on opposing hills about 1,000 yards from U.N. lines, and the 780th was called in to see what they could do. They mounted one of their howitzers on a 155-mm chassis and towed, pushed and drove this combination up a mountain to an altitude of about 3,000 feet. The crew literally bore-sighted the howitzer and destroyed 20 of the bunkers on the first day of firing.

The Communists again took advantage of the talks to build up a force of nearly one-million Chinese troops in North Korea by early 1953. Then, they attacked in an attempt to end the war with a military victory. Battles raged back and forth across the cease-fire line through June and early July 1953, when it became obvious to the Communists that a military victory would be denied them. The Marine's 11th Field Artillery continued their fire support in the mountains near the 38th parallel during the negotiated cease fire, and defended the Demilitarized Zone from the end of hostilities in 1953 to March 1955.

On July 27, an armistice was signed by the belligerents and the fighting ended. No one won, no one surrendered. But Communist expansion was stopped so the U.N. goals were realized. This limited war cost six-million dead

"Cannoneers" by Joyce Kreafle

First Marines launch a 4.5-inch rocket barrage against the Chinese Communists. (Photo courtesy of the National Archives)

A 155-mm howitzer gun crew of the 25th Infantry Division fires on North Koreans near Uirson. (Photo courtesy of the National Archives)

Battery B, 101st Field Artillery Battalion, 3rd Infantry Division, fires a 105-mm near the Han River. (Photo courtesy of the National Archives)

A self-propelled 8-inch howitzer of the 780th Field Artillery in action in January 1952. (Photo courtesy of George J. Ellis)

A gun crew of the 2nd Infantry Division digs in for action against Korean forces near the front lines on August 27, 1950. (Photo courtesy of the National Archives)

and wounded on the two sides, about half of this number being South Korean civilians. U.N. forces suffered 472,344 killed, of which 54,246 were Americans.

Artillery tactics during North Korea's initial drive south were similar to those learned during jungle fighting in the Pacific in World War II, although Korea (North or South) was hardly a jungle. Guns and howitzers were elevated for firing on close-range targets, and had to be protected in all directions against infiltrating enemy infantry. During the pursuit northward in the fall of 1950, however, artillery could be used in a more conventional way to support the U.N. infantry attacks and shell the retreating enemy. In the Chinese drive south in the winter of 1950-51, some artillery units of the 2nd Division lost many of their pieces in this forced withdrawal towards the 38th parallel. The 37th Field Artillery in a rear-guard action lost ten artillery pieces to the enemy.

When the fighting settled down near the 38th parallel to what was essentially World War I trench warfare with World War II equipment, artillery became important to both sides. Thunderous artillery barrages ensued. Massed fires of 20 artillery battalions were not uncommon across the mountainous front lines. One time, a million rounds were fired in three weeks from the artillery pieces of the American 2nd Division. Such barrages stifled all attempts of the enemy to advance from his positions. Together with air and naval bombardments that could be called in at any time, massed artillery fires ended all enemy attacks. Still, artillerymen wanted more firepower. For Korea, however, additional firepower did not materialize as the stalemate fostered negotiations that still continue.

Two tired men of a gun crew of the 37th Field Artillery Battalion, 2nd Infantry Division, sleep as the rest of the crew fires on Korea on March 18, 1951. (Photo courtesy of the National Archives)

An 8-inch self-propelled howitzer preparing for direct assault fire on Communist positions on January 17, 1951. (Photo courtesy of the National Archives)

A 155-mm howitzer mounted on an M-41 chassis, 92nd Field Artillery Battalion, Eighth Army, moves into a new position on February 24, 1952. (Photo courtesy of the National Archives)

Post-Korea Artillery Developments 1953–1962

Following the end of the Korean War, the Marines were called on for occasional overseas duty. In July 1958, the 10th Marines (Artillery) accompanied a Marine contingent to Lebanon to quell fighting between warring factions in that unhappy country. In the 10th's inventory of weapons were 105-mm howitzers, 8-inch howitzers and 4.2-inch mortars. The 10th was withdrawn from Lebanon in October 1958. In April 1965, part of the 10th Marines accompanied an incursion into the Dominican Republic to put down an attempted Communist coup. Three artillerymen were ambushed and killed. An international peacekeeping force relieved the Marines in June 1965.

During and following the Korean War, the Army developed new artillery pieces, some adopted, some not: M52 105-mm, M44 155-mm and M55 8-inch howitzers, and the M53 155-mm gun. All were self-propelled, but none was capable of 360 degree traverse or high-angle (75 degrees) firing. None was used in combat. Also developed and deployed in 1955, were the 80-mile-range "Corporal" guided missile and the 30-kilometer-range "Honest John" rocket; both could be armed with a nuclear warhead. The 280-mm nuclear shell-firing gun, "Atomic Annie," was developed during the Korean War, but was too heavy and unmaneuverable and had a disappointing range of 25 kilometers. It was generally unsatisfactory as tube artillery. Only one nuclear shell was ever test-fired from the big gun. About 20 of these were made and shipped to Germany during the height of the Cold War with the Soviet Union. But unprecedented — virtually unlimited — firepower was now available to commanders on the battlefield. However, debate still raged over how, or if, to employ these weapons of mass destruction. There also arose interservice ri-

Men of the 320th Airborne Field Artillery Battalion before a parachute drop at Fort Campbell, Kentucky, in 1954. (Photo courtesy of the 320th Airborne Field Artillery Battalion Association)

Parachute delivery of 105-mm howitzers. (Photo courtesy of the 320th Airborne Field Artillery Battalion Association)

De-rigging of 105-mm howitzers on drop zone after parachute delivery. (Photo courtesy of the 320th Airborne Field Artillery Battalion Association)

valry between the Army, Air Force and Navy over the control and use of nuclear weapons. This was a two-way street: the enemy, the Soviet Union at this time, had developed nuclear weapons also.

The nuclear battlefield forced the Army to rethink offensive and defensive tactics and develop the concept of the battle group, a small, highly mobile, dispersed unit of a division, relatively self-supporting, and with its attached artillery. New weapons were needed, ones light enough to be airborne, could traverse 360 degrees, and could replace the cumbersome first-generation nuclear missiles. These needs developed by 1956, the pentomic infantry division, a unit of five semi-independent battle groups, each with five rifle companies, one support company (with a mortar battery), all supported by armor, cavalry, engineer and artillery battalions. A battle group was between a battalion and a regiment in size. The artillery component was composed of one 105-mm howitzer battalion (five batteries, 30 towed pieces), one 155-mm howitzer battalion (12 pieces), one 8-inch howitzer battery (4 pieces) and an Honest John rocket battery of two launchers. The artillery battalions, 155-mm and up, were nuclear-capable. The total of nine batteries per division had a complement of 1,763 men. Similarly, airborne divisions were formed into pentomic divisions, although with fewer artillery batteries. Armored divisions retained their three combat commands, their armored infantry and tank structure (four battalions each) and added a composite artillery battalion of 155-mm and 8-inch howitzers and Honest John launchers. All the new artillery tactics were based on the use of nuclear warheads on the battlefield, and this and the pentomic concept, itself, were not without critics. By 1960, the army's nuclear artillery was large, cumbersome, unmaneuverable and too heavy for airlifting, and its conventional weapons still of World War II vintage.

The pentomic concept did not last long. By 1962, there was developed the Reorganization Objective Army Divisions (ROAD), which adopted four division types: Infantry, Armored, Airborne and Mechanized. All had a division headquarters, three brigade headquarters (the number of battalions per brigade varied), support, aviation, engineer and signal battalions, a cavalry squadron and a military police (MP) company. The new organization was very flexible according to the mission.

This new organization was a return to the triangular regimental system of World War II, but the ROAD brigade was very different. It had a flexible headquarters with battalions attached according to the mission. An infantry division had eight infantry battalions and two armored battalions. The armored division possessed six armored battalions and five mechanized infantry battalions. A mechanized infantry division consisted of three tank battalions and seven or more mechanized infantry battalions, and an airborne division had nine airborne infantry battalions. The division commander could tailor his command by attaching battalions to the brigades to fit a particular situation or mission.

Also in the 1960s, new or improved and redesigned towed and self-propelled weapons appeared, (self-propelled M108 105-mm, M109 155-mm, and M110 8-inch howitzers and towed M102 105-mm, M114 155-mm and M115 8-inch howitzers) among them the self-propelled M107 175-mm gun. New Little John (10-mile range), Sergeant rockets (93-mile range), and the Pershing I missile (up to 500-mile range) were accepted. Some were too much vehicle for a relatively small gun, and some were based on technology of the 1920s. Helicopters were fitted with rockets and a new computerized firing system was adopted (Field Artillery Digital Automatic Computer, or FADAC). Perhaps the most significant development in these years was the Air Assault Division with 12 assault helicopters plus infantry, artillery with 105-mm howitzers and "Little John" rockets and support units, all transported by helicopter.

"Fire Base Vietnam" by Joyce Kreafle

Vietnam War 1956–1975

North and South Vietnam combined are only slightly larger than the state of New Mexico in land area, but with a total coastline on the South China Sea of about 2,000 miles. The United States' involvement in the Vietnam conflict (war was never officially declared) was slow at first, then it escalated dramatically, and eventually became a quagmire — a sinkhole of men, materiel and money — from which American forces were eventually withdrawn. In 1956, the U.S. stepped in following the defeat of the French at Dien Bien Phu in May of 1954, by the communist Viet Minh forces of Vietnam's leader, Ho Chi Minh.

Severe restrictions hobbled America's fighting men in South Vietnam. Among these were the eventual cessation of bombing of North Vietnam (the source of the enemy's military supplies). A decision at the highest levels in Washington prohibited the pursuit of retreating enemy into "neutral" Cambodia and Laos, which he used as a sanctuary when battles were going badly for him. These restrictions made this unpopular war virtually impossible to win.

Conflict in French Indochina (Vietnam) dates from 1947, when guerrillas, under the control of Communist Ho Chi Minh and the French fought for control of the country. The French were beaten in 1954, and withdrew from the country, as noted above. The Geneva Conferences of that year partitioned the area into North (Communist) and South Vietnam. Although he accepted the partition, Ho Chi Minh's goal was to overthrow the South Vietnamese government and to eventually unite the two countries into a single Communist State. The Army of the Republic of Vietnam (South Vietnam), ARVN, outnumbered the 20,000-man Communist Viet Cong (VC) guerrilla force ten to one, but the ARVN was unable to protect the villages in the swamplands and jungles from VC raids. Much war materiel and many troops to aid the VC streamed down from Communist North Vietnam, including the VC's mainstay artillery weapon, the easily portable 120-mm rocket launcher. The government of South Vietnam formally asked the U.S. for aid to prevent the collapse of its government. By 1961, the U.S. had about 750 military "advisors" in South Vietnam, supposedly noncombatants, but air units were already flying missions to help the ARVN. By early 1962, American presence in South Vietnam numbered about 2,700 and by late 1963, more than 16,000 men.

On July 30, 1964, North Vietnamese torpedo boats, while pursuing South Vietnamese raiding Navy boats, attacked two U.S. Navy destroyers in the Gulf of Tonkin. Two of the torpedo boats were sunk and their bases bombed from the air. President Lyndon B. Johnson, in office less than a year and with full support of the U.S. Congress, regarded this incident as a legal basis for a dramatic increase in American support for the South Vietnamese government. The war suddenly escalated. Large numbers of North Vietnamese Regular Army (NVA) troops were sent into South Vietnam from the North in the Spring of 1965. Two American camps at Pleiku in the Central Highlands were attacked, and the U.S. Air Force began bombing North Vietnam from land bases and from aircraft carriers. Before the U.S. disengaged itself from this conflict, its Army, Navy and Marine air forces had dropped three times the tonnage of bombs on the two Vietnams as had been dropped in both the European and the Pacific theaters of war in World War II.

The 12th Marines' artillery was sent to Vietnam in May 1965, and there saw action at Da Nang, Chu Lai, Phu Bai and Hue. The following January, the 11th Marines arrived with its artillery to support the 1st and 3rd Marine Divisions fighting at Quang Tri, Thua Thien and An Hoa in addition to those noted above for the 12th Marines. Many of these engagements were the jungle fighting of World War II Pacific islands, while others were in defense of populated areas under attack by the Viet Cong. Few major engagements were fought in Vietnam without artillery support, and artillery was critical in the defense of isolated fire bases.

With the need for more manpower in South Vietnam, the U.S. mobilized Army, Navy and Marine units in 1965, and by the end of the year, tens of thousands of Army soldiers and Marines were in the jungles. By the middle of 1969, there would be more than a half-million U.S. troops in South Vietnam. The Vietnam War was a different kind of fighting for all ground troops. There were no front lines and no major, concrete objectives as, for example, the breaching the Siegfried Line in Germany or taking Mount Suribachi on Iwo Jima in World War II. The Army and Marines established defensive perimeters around fire bases, cities, and airfields, and deployed on search-and-clear missions into the jungle (frequently flown in by helicopter) against reported Viet Cong and North Vietnamese troop concentrations. Artillery pieces and ammunition were flown in by CH47 Chinook helicopters. The first such airmobile unit was the 11th Air Assault Division (towed 105-mm howitzers).

For the "grunts" and "dogfaces," the next four years would be search-and-destroy, search-and-clear, search-and-seize, and cordon-and-search. Often, contact was made with the enemy — sometimes not. Jungle battles were often fought at platoon and company strength levels, occasionally at battalion strength. The NVA and Viet Cong never allowed masses of its troops to be engaged in a single battle for fear of total annihilation. Theirs was a war of attrition. A protracted conflict of 10, 15 or 20 years was of no consequence to their leaders. This was a different kind of war.

Firing to the north into II Corps from fire base Joannie. (Photo courtesy of Robert W. Flournoy)

Battery B, 5th Battalion, 42nd Field Artillery, firing a 155-mm for close support of the 1st Cavalry. (Photo courtesy of Robert W. Flournoy)

A gunner from the 25th Infantry Division, throws a brass casing into a stock pile after the 105-mm howitzer is fired at fire support base "Wire" October 4, 1968. (Photo courtesy of the National Archives)

Brigadier General D. S. Henderson, Assistant Division Commander, 25th Infantry Division, fires the one-millionth round expended by the 1st Battalion, 8th Field Artillery, in Vietnam. (Photo courtesy of Headquarters, 25th Infantry Division (Light) Artillery)

In the mid-1960s, the need for skilled field artillery officers for duty in Vietnam finally caused the Army to reorganize its artillery. At this time, artillerymen were trained in both Field Artillery and Air Defense tactics and procedures. The one-year duty tour in the battle zone of Vietnam left little time for on-the-job training in either specialty. Combat in Vietnam required the officer to arrive as a proficient field artilleryman and not a hybrid field and air defense artilleryman. Army commanders in Vietnam did not have time to train an air defense artilleryman to be a competent field artilleryman, nor to upgrade the skills of a field artilleryman, who had insufficient training in the basics. A study urged ending the practice of cross-training and recommended forming two separate branches of artillery.

The Army accepted the recommendations and, in 1968, separated field and air defense artillery into two independent combat arms with their own training schools. This freed field artillerymen to concentrate on field artillery subjects. The separation to two branches had little impact on the Artillery and Guided Missile School, however, which was renamed the U.S. Army Field Artillery School because its emphasis was already on field artillery matters.

The war in Vietnam continued into the 1970s and became more and more unpopular in the United States. Troops started pulling out of Vietnam in the late 1960s, and American troop strength in Vietnam was down to 175,000 by the beginning of 1972. The NVA invaded South Vietnam in March 1972, by simply and boldly marching across the Demilitarized Zone at the 17th parallel. Several rounds of negotiations between the U.S. and the two Vietnams broke down until finally, in January 1973, peace accords were signed in Paris, and the U.S. began to plan its final withdrawal from South Vietnam. In January 1975, North Vietnam began its final push to defeat the ARVN forces in South Vietnam, and it went faster than planned. By April 30, 1975, the NVA was in Saigon and the South

E Battery, 1st Battalion, 5th Field Artillery, at Fire Support Base ORAN in the Republic of Vietnam in February 1969. The 4.2-inch mortars, assigned to the infantry units, were routinely attached to the artillery. Known as "Light Horse," they were extremely advantageous because they could be airlifted with crew by the UH-1 Huey helicopters, rather than the CH-47s necessary to move 105-mm howitzers. (Photo courtesy of the Headquarters, 1st Battalion, 5th Field Artillery)

Gun crew awaits firing orders. (Photo courtesy of the National Archives)

A self-propelled 155-mm howitzer during Operation Dewey Canyon II fire support. (Photo courtesy of the National Archives)

surrendered unconditionally to the North. The last U.S. forces were evacuated from Saigon by the end of April, and America's 20-year involvement in the land war in Southeast Asia ended.

The two Vietnams became one country in 1976, under Communist rule. Although American forces never lost a major battle in this war, the final cost in lives and money was overwhelming. More than 58,000 Americans were dead, 304,000 American wounded, and 150 billion dollars expended. Two to three million North and South Vietnamese were estimated killed. The casualty rate of the 2.6 million American Vietnam veterans was a staggering 10 percent. The fate of Americans missing in action (originally some 2,500 names) remains unresolved at this writing, and the U.S. still does not recognize the government of Vietnam.

Helicopters and aerial artillery came into their own in the Vietnam war. These rotary-wing aircraft carried troops into disputed jungle and mountain regions and later evacuated them after their mission was completed. Artillery was airlifted into fire bases, and heavily armed helicopter gunships wreaked havoc on attacking enemy troops, then they acted as air observers to adjust tube-artillery fire on the ground. The airmobile concept proved itself early in the war. For artillery on the ground, it was necessary to position batteries close enough together to be mutually supporting against attacks from all directions. But there were disadvantages: the batteries could not be moved through the jungle and often battles were out of range. The only solution for increased firepower to help the troops on the ground, in these cases, was the helicopter gunship, in essence, a flying artillery.

The 1st Cavalry soon learned to move 105-mm and 155-mm howitzers anywhere in the country where a fight needed them. And some of these new weapons, like the M102, M108 and M109 howitzers, could traverse 6,400 mils (360 degrees), a necessity in guerrilla warfare. This was a small-unit war, and isolated batteries soon learned to perform like battalions, such that critical decisions in the field were now those of the battery commander. FADAC was being used extensively by 1969, and it worked extremely well. In the early 1970s, the Army began its "Vietnamization" programs, turning over the war completely to its South Vietnam allies. It did not work and eventually invited a disaster.

Post-Vietnam Artillery Developments 1976–1990

With the Vietnam war now in the past, the Army looked again at its primary military rival, the Soviet Union, and saw an enormously powerful adversary that had become the leading power in Europe. Artillery began to deploy, in the late 1970s, a replacement for FADAC. The new system was called Tactical Fire Direction System (TACFIRE). This system was designed to reduce the time between a request for fire and the first round on target, and to cut the time required to complete fire planning calculation from hours to seconds. The observer communicated directly with the computer at the fire direction center using a message entry device. After verification, the computer entered all pertinent data and singled out the battery, or batteries, in the best position to fire on the target. If the target was beyond the battalion's capability, the computer automatically requested action from a higher headquarters.

The first units to be equipped with TACFIRE were one battalion of the 17th Field Artillery and the 1st Cavalry Division, but even by 1980, few artillerymen knew anything of its great potential. There was little doctrine on the new system. Also, two new radar systems were developed, one that located and tracked fire of enemy mortar batteries (AN/TPQ-36) and another that did the same against enemy gun/howitzer batteries (AN/TPQ-37). Both did the computations and fed information of incoming projectile trajectories into the TACFIRE computers. Also, field artillery tested the Remotely Piloted Vehicle (RPV), which was controlled from the ground, flown over enemy territory and acquired potential targets by television and laser designators for strikes by precision-guided munitions. The RPV program developed into the Aquila RPV in the late 1970s. Precision-guided projectiles, or so-called "smart" munitions, could be laser guided by an observer on the ground (or by manned aircraft or by RPV) who would "paint" a visible target with a laser beam, and the laser-sensing projectile would almost invariably score a direct hit, whether the target was a single moving tank or the front door of a building. The Copperhead, for example, was a fin-stabilized projectile fired from a 155-mm tube with such laser-guided capabilities.

Still more developments were pushed in tube-artillery design in the 1970s and 1980s. Rocket-assisted projectiles from towed M198 155-mm howitzers had a range of 18 miles. Honest John and Sergeant missiles were replaced with the Lance missile in the 1970s, and at this time, there began the design of a rocket system called the General Support Rocket System, later to become Multiple Launch Rocket System (MLRS). It was to be a counterbattery and enemy air defense suppression weapon. All these developments enormously increased the amount of artillery firepower, and reduced reliance on using nuclear weapons on a European battlefield. Also, reorganization of forward observer duties was undertaken in the late 1970s.

The size of the modern battlefield had enlarged greatly, and the forward observer team now could not supply observed fire support throughout its company's sector. The Army also required the capability to rapidly shift weapons on the battlefield, and mass fire from several weapons systems — mortars, field artillery, attack helicopters and tactical aircraft. With this objective in mind, it was proposed to reorganize the forward observer team to take better advantage of the new technology appearing and scheduled to be in the Army's inventory in the 1980s. To accomplish the objective of enhancing forward observation, a study group urged creation of a Fire Support Team (FIST) at company level to replace the forward observer team. The FIST was critical, the study said, because the artillery had fewer pieces than the enemy; because the envisioned battlefield would be larger than in the past; because the forward observer had a major role in registrations and the use of smoke and illuminations that the fire direction center had previously coordinated; because new equipment and new munitions were being introduced; and because airborne, forward tactical air-controllers would likely be absent because of actions by Soviet-bloc air defenses.

One of the significant doctrinal changes was to give each division a field artillery brigade in a reinforcing role. This placed the positioning and fire control of corps artillery cannon battalions under division commander's authority and assured all cannons in a division sector would be completely responsive to division artilleries. Along with this change, the responsibility for counterbattery fire was given to division artilleries, and the Q-37 counterbattery radar was made organic to division artilleries.

The FIST concept was not without its opponents, but was eventually adopted, with refinements, by the field artillery. There evolved in the mid-1970s the concept of active defense, the idea that against an attacking opponent overwhelming in numbers, enormous doses of firepower had to be applied during the first hours of defense and before counterattacking. The artillery, tube and missile, was central to the success of this doctrine.

Further, for implementation in the 1980s, a study to completely restructure the division was undertaken to eventually replace the ROAD structure of the 1960s. One truly innovative suggestion was the creation of combat units around the new weapons rather than fitting new weapons into existing divisions and smaller units. The heavy division with more battalions was envisioned as a major battle unit and field artillery in this division was strengthened considerably. This heavy division was approved in 1980; its artillery consisted of 72 155-mm (M109 self-propelled) and 16 8-inch howitzers (M110 self-propelled) plus nine support rocket systems. As part of this new heavy division there were five or six tank battalions and four or five mechanized infantry battalions, the numbers of the tank and mechanized battalions presumably depended on the mission. The 105-mm howitzers had long been dropped from the heavy division's inventory in favor of heavier weapons.

In addition to introducing the heavy division, the Army in the 1980s studied and developed the light divisions, after many changes and false starts. This light division could be quickly airlifted anywhere in the world on short notice, to fight low- or mid-intensity battles as opposed to the armored, nuclear conflict envisioned against the Soviet Union in Europe. Artillery in this light division or Contingency Force was a mix of eight M198 155-mm howitzers and 54 M102 105-mm howitzers, which were later replaced with a lightweight, British-designed 105-mm piece and redesignated M119. It was a trade-off between firepower and mobility.

In the late 1970s and early 1980s, the Soviet Union not only modernized its weapons, but also began to echelon its forces, that is, establish a behind-the-front, deep second and third striking force that would be a follow-on to the initial wave. This gave another and imperative mission to artillery to not only confront the first wave of attacks, but also disrupt the deep, echeloned forces. These new "deep battle" tactics became known as AirLand Battle doctrine and by 1980, replaced the "active defense" concept of the mid-1970s. Counterbattery and countermortar radars were ordered improved, resulting in the Firefinder Radar System and Ground/Vehicular Laser Locator Designator (G/VLLD). OH-58D helicopters were assigned to the field artillery as a result of the 1981 Army Helicopter Improvement Program (AHIP) and carried laser-target designators that would give a variety of artillery "smart" projectiles pinpoint accuracy on first firing. By the end of the decade, however, artillery would lose its priority for the helicopters. Ground-fired missiles also received considerable attention. The MLRS, fielded in 1983, was to be the launcher for the Army Tactical Missile System (ATACMS) and was to replace the Lance missile in the 1990s, and have a longer range than the Lance's, among other improvements. Disarmament developments however, slowed down the program. The M109 self-propelled 155-mm howitzer, fielded in 1963, was updated many times over many years by the Howitzer Improvement Program (HIP) and in 1991, the Army fielded the latest model, the M109A6 Paladin. This new system had on-board navigation and fire control capabilities so each howitzer could stop, know its location and deliver fire on an assigned target — a major improvement in survivability. New fire control systems were studied in the 1980s, including the Advanced Field Artillery Tactical Data System (AFATDS), which would replace TACFIRE. Restructuring increased the number of tubes per battalion to 24, from the six-gun battery to the eight-gun battery, and MLRS batteries were increased. Also deployed in the early 1980s was the highly accurate Pershing II missile with a range of more than 1,000 miles. It was withdrawn in 1988 by treaty agreements. Finally, the number of observation laser teams was increased in the heavy division to take advantage of the increasing number and types of "smart" munitions.

Grenada

Field artillery was called upon during the Cold War years of the 1980s on three different occasions. On October 25, 1983, the island of Grenada, southernmost of the Caribbean Windward Islands, was the subject of an incursion by a U.S. task force in part to forestall a Cuban/Soviet Union military buildup on this Marxist-led island nation. At first, General H. Norman Schwarzkopf was an advisor, then later the Task Force Deputy Commander. Artillerymen of the 82nd Airborne Division and a battery of the 3rd Battalion, 10th Marines, provided effective fire support for the Joint Task Force, one of whose missions was to rescue American citizens. The operation was highly successful. All U.S. troops were withdrawn from Grenada in December.

Lebanon

The 10th Marines supported the multinational peacekeeping force in Lebanon from 1982 to 1984. Several 3rd Battalion Marines were killed in a bomb explosion at the airport on October 23, 1983. Target acquisition units from Fort Sill, Oklahoma, were sent to Lebanon in 1983 in support of U.S. Marines deployed there. These units functioned in a counterbattery role, using advanced radar equipment.

Panama

Security forces from the 2nd Marine Division deployed about 1,300 men to Panama in April 1988. Later 800 more Marines shipped out to Panama for jungle training. With these was a forward observer team from the 10th Marines. On December 20, 1989, U.S. Task Forces were flown into Panama to quell anti-U.S. riots staged by the dictatorship. This followed a failed October coup attempt to overthrow the corrupt government. This was called "Operation Just Cause," and it lasted four days. Seven thousand U.S. troops were flown into Panama on the first day, followed by 2,500 the next day, to reinforce 12,700 Army, Marine, Navy and Air Force troops already there. Strongman Manuel Noriega, indicted on drug trafficking charges by U.S. courts and threatening a treaty-breaking takeover of the Panama Canal, was deposed and taken into custody. Twenty-three U.S. soldiers were killed in action and 322 were wounded. When fighting started on December 20 in Panama, the Marines immediately seized the Bridge of the Americas, the only direct route into Panama City across the canal from Rio Hato, a Panamanian armory.

Although most American forces deployed were light infantry, airborne and Ranger units, supporting light artillery played a key role by, in one instance, reducing the headquarters of the Panama Defense Forces in Panama City to shambles with devastating direct fires of howitzers.

Elements of the 2nd Battalion, 8th Field Artillery, commanded by Lieutenant Colonel Perry F. Baltimore III, were deployed with Ready Brigade elements of the 7th Infantry Division (Light) to Fort Sherman, Panama, in May 1989. Their mission was to protect American families in the Caribbean end of the Canal Zone, which was accomplished without incident. In October 1989, these forces were relieved by the 7th Division's 3rd Infantry Brigade with the 7th Battalion, 15th Field Artillery, as the fire support section. On December 20, the 7th Division's 2nd Brigade with the 6th Battalion, 8th Field Artillery (18 M102 105-mm howitzers plus a Q-36 radar section), arrived in full complement. The 2nd Battalion, 8th Field Artillery, was recalled to Panama at this time also, but without howitzers. The rules of engagement in Panama were very restrictive. Armed force was the last resort in all operations, and harm to civilians was especially avoided. Early successes limited the use of artillery in the Panama operations to mostly isolated show-of-force missions at road blocks, for example, except for the destruction of the Panamanian Defense Forces headquarters building noted above.

Virtually all of the changes in field artillery equipment, technology and tactics since the end of World War II had taken place with one idea in mind, to blunt a Soviet Union attack on Europe and eventually defeat its enormous military machine on the battlefield, and its deeper echeloned forces, with conventional or nuclear weapons — whatever it took. The attack never came. Negotiations with the Warsaw Pact nations in early 1989, including the Soviet Union, brought about agreements in force reductions on both sides. The Warsaw Pact dissolved in 1990, the two Germanys were united, and in 1991, the Soviet Union, itself, disintegrated as a monolithic state. Almost overnight, the great threat of war in Europe seemingly evaporated. U.S. artillery programs were cut back, and with forced reduction in personnel, new technology now became more important than ever to maintain a firepower edge. But in looking back over the 1970s and 1980s, the artillery modernization programs showed only minimal substantive results. Only four systems had been added over the 20 or so years: the M198 155-mm howitzer, the MLRS, the Q-36 and Q-37 counterfire radar systems and the TACFIRE. These systems would now be tested in battle.

Persian Gulf War 1990–1991

Saddam Hussein, the volatile dictator of Iraq, had barely extracted his country from an eight-year war with neighboring Iran when he attacked Kuwait, his small neighbor to the south. The goal of Hussein in both conflicts was territorial expansion and complete control of the waterway at the head of the Persian Gulf. This waterway was Iraq's only outlet to the sea to ship its oil by tanker. After making impossible demands on Kuwait, Iraq broke off diplomatic relations on August 1, 1990, and invaded Kuwait with 100,000 troops the next day. Hussein had one-million men under arms, at least 5,500 tanks, 3,500 artillery pieces and about 700 aircraft. Kuwait's army consisted of 20,000 men. Complete occupation of Kuwait by Iraq's far superior forces and equipment (mostly of Soviet-bloc manufacture) took barely hours. In the next few months, Kuwait would be ravaged and robbed by Iraqi troops, and anything of value was shipped back to Baghdad, the Iraqi capital. Before they were finally forced out six months later, Iraqi troops set fire to some 600 of Kuwait's 1,000 oil wells and released millions of gallons of crude oil into the Gulf. Following Iraq's occupation of Kuwait in August 1990, the Army's 82d Airborne and Marines were rushed to Saudi Arabia as a first line of defense should Iraq decide to push into that country. Eventually, the 10th, 11th, and part of the 12th Marines (Artillery) would be in Saudi Arabia and would participate in the ground-war phase of operations in Iraq and Kuwait.

Desert Shield 1990

The United Nations, led by the United States, denounced the invasion and imposed sanctions on Iraq, including shutting down Iraq's overland oil pipelines through Turkey and Saudi Arabia. Intelligence data indicated Hussein's next victim would be Saudi Arabia. The United States, with Saudi permission, began the buildup of ground and air forces in Saudi Arabia and naval forces in the Gulf. If Iraq were permitted to overrun Saudi Arabia and other small neighboring oil-rich states, it would then control more than 40 percent of the world's known oil reserves. The entire U.N. buildup operation was called "Desert Shield" and, at full force, involved more than 40 nations, large and small (the Coalition Forces). U.S. forces alone, totaled more than a half-million men and women, more than 100 ships, including two battleships and five modern aircraft carriers, in addition to missile cruisers and many smaller service and support vessels. The U.S. had 1,800 fixed wing aircraft and an almost equal number of helicopters, and 2,000 tanks and other

"Desert Thunder" by Joyce Kreafle

armored vehicles. U.S. ground forces included the 1st and 2nd Marine Divisions, the 1st Brigade of the 2nd Armored Division, 1st Cavalry Division, the 2nd and 3rd Armored Cavalry Regiments, the 1st and 3rd Armored Divisions, the 1st and 24th Infantry Divisions (Mechanized), the 82nd Airborne and the 101st Airborne (Air Assault) Divisions. The U.S. forces were organized into two Corps, the XVIII Airborne and the VII Corps. Also at the front were the French 6th Armored (Light) Division, the British 1st Armored Division, an Egyptian armored corps, a Syrian infantry division (mechanized), a Saudi infantry division (mechanized) and a Saudi National Guard infantry task force (mechanized).

At the time Desert Storm (the ground war) began in February 1991, American field artillery had completed deployment of a massive force. The Army artillery force consisted of 43 cannon, rocket and missile battalions organized into two Corps Artilleries, seven Division Artilleries and seven Field Artillery Brigades. Two corps artillery headquarters provided overall command and control. The artillery pieces available were 108 M102 (105-mm, towed) Howitzers, 642 155-mm Howitzers (towed and self-propelled), 96 M110 (203-mm) Howitzers and 189 Multiple Launch Rocket Systems. This last included 18 ATACMS-capable launchers. A Field Artillery Brigade's logistical requirements are mind-boggling. The 75th Field Artillery Brigade, for example, with more than 2,000 men and 634 tracked and wheeled vehicles (and 195 trailers), required — daily — 500 cases of meals ready-to-eat (MRE), 10,000 gallons of bulk water, 500 cases of bottled water and 30,000 gallons of bulk fuel, if on the move (5,000 gallons, if stationary). The 75th Field Artillery Brigade supported both XVIII Airborne and VII Corps during Desert Storm.

The first airlift by a UH-60L Black Hawk helicopter of a 155-mm towed howitzer. The howitzer is the world's first 30km-range, 155-mm towed howitzer that weighs less than 9,000 pounds. (Photo courtesy of Francis J. Hoerster)

Desert Storm 1991

More than five months of negotiations ensued, but Iraq would not withdraw from Kuwait. A January 15, 1991, deadline was imposed by the U.N. Coalition, but Iraq showed no evidence of withdrawing from now-ravaged Kuwait. On January 17, the Coalition struck. This was the beginning of Operation Desert Storm. Massive strikes by land- and carrier-based fighter-bombers and Cruise missiles (100 missles in the first wave) were launched at military targets in Iraq. Iraq's response was the launching of Scud missiles into Saudi Arabia and Israel. These Scuds were unguided terror weapons. In less than one week, the Coalition had achieved absolute air superiority, and many aircraft of what was left of Iraq's air force (half was destroyed) had fled to Iran. Virtually all its communications from Central Headquarters to its units in the field were bombed and disrupted. All bridges and roads useful to its troops were destroyed, and Iraq's military transportation and supply systems came to a halt. The air war lasted 40 days and reached 3,000 sorties per day.

Toward the end of January, about 1,500 Iraqi soldiers with 100 tanks and armored vehicles invaded and occupied the abandoned Saudi town of Khafji just south of the Kuwait border. The next day the Marines helped a multinational force retake the town in house-to-house fighting. An Iraqi amphibious relief force was destroyed at sea by air attack. Meanwhile, offshore, a 17,000-man Marine amphibious force circled in the Gulf off Kuwait, threatening an assault. It was a feint to tie down six Iraqi divisions, which it did.

February 23, 1991, was the new deadline for Iraq's acceptance of the U.N. resolutions of withdrawal if it were to avoid a land war. The deadline came and went, and on February 24, G-Day, General Schwarzkopf sent his Coalition troops and armor into Iraq and Kuwait in a massive and overwhelming assault. Coalition forces swept into Iraq and Kuwait from Saudi

Gun #5 of "Echo" Battery, 3rd Battalion, 11th Marines, shooting a rap round on first day of the ground war. (Photo courtesy of Headquarters Battery, 11th Marine Division)

Arabia in a huge, high-speed, clockwise attack, first north then east, trapping and annihilating Iraqi forces as they attempted to escape northward from Kuwait. This ground-attack phase of Desert Storm lasted 4 days. All of Iraq's military forces in the War Zone were destroyed or rendered ineffective.

Some units were in action before this, however; the 3rd Battalion, 82nd Field Artillery, and the 2nd Brigade of the 1st Cavalry Division had already fired in support of an attack near Kuwait's western border with Iraq to test Iraqi responses. (The 3rd Battalion would fire 3,529 rounds during this short war.) Iraqi artillery and air defenses were the targets to open safe corridors for Coalition attack helicopters. When the Iraqis turned on their counterbattery radar, it was promptly located and destroyed. The 5th Battalion, 11th Marines (Artillery), 1st Marine Division, shelled four enemy positions inside Kuwait between January 23 and February 10, 1991. Their ordnance included M109s (155-mm, self-propelled), M198s (155-mm, towed) and M110s (203-mm, self-propelled). One target was an Iraqi battery, 20 kilometers (12.4 miles) inside Kuwait. All positions were put out

Soldiers, howitzers, and other equipment of the 1st Battalion, 5th Field Artillery, in April 1991, in Area "Allen" in southern Iraq following the cease fire in Southwest Asia. During Operation Desert Storm, the unit provided fires in direct support of the 1st Brigade, 1st Infantry Division (Mechanized). (Photo courtesy of Headquarters, 1st Battalion, 5th Field Artillery)

of operation by 5th Battalion, 11th Marines' fires, aided in one instance by a Marine F/A-18 fighter-bomber. Iraqi counterbattery fire was almost nonexistent.

On G-Day, the 1st Infantry Division (U.S.), the 1st Cavalry Division (U.S.) and the British 1st Armored Division attacked from Saudi Arabia into Iraq, just west of the Kuwaiti border. The border with Saudi Arabia is marked by a berm that had to be breached. During this breaching operation, three Field Artillery Brigades and two Division Artilleries supported the 1st Infantry Division's attack with 6,000 cannon rounds and more than 400 rockets. With such overwhelming firepower, there was no resistance and no enemy counterfire.

The 1st Armored Division Artillery, near the middle of the Coalition front, delivered 1,213 rockets and more than 9,500 rounds of cannon fire in its 218-mile drive into Iraq and Kuwait. In this drive north and east into western Kuwait, the 1st Armored Division Artillery employed the MLRS as the system of choice in counterfire work, silencing all enemy batteries that fired at them. Although two Iraqi rocket systems and four cannon systems out-ranged the MLRS, the enemy lacked a reliable target acquisition system, and so their whole response effort was compromised. The 1st Armored Division Artillery M109 howitzers were also outranged by the Iraqi cannon and rocket systems, but again, superior target acquisition systems and especially the training of U.S. artillerymen, countered this potential enemy advantage.

On the far right, Marines of the 1st and 2nd Divisions with a Pan-Arab force attacked into southern Kuwait when the ground war began in February and proceeded to encircle the country. At the Kuwait International Airport, the Marines ran across a 100-tank Iraqi armored force and promptly destroyed every tank.

On the far left, the XVIII Airborne Corps with three field artillery brigades (the 18th, 196th and 212th) drove north with the French 6th Armored Division (Light) and the U.S. 24th

Infantry Division, into Iraq. Between February 13 and 23, before G-Day, these brigades fired in support of deep battle operations of the 24th Infantry Division on their right and the 82nd Airborne Division on the far left. On the final day of Desert Storm, February 28, the 18th and 212th Field Artillery Brigades and 24th Division Artillery massed nine battalions near Basra, Iraq, in an early morning barrage that destroyed the Hammurabai Republican Guard Forces Armored Division.

After 100 hours of fighting, it was all over. Iraqi troops were fleeing the armored onslaught in total disorder. Some of their commanders had deserted them. U.S. President George W. Bush called for a cease fire, but the U.N. goals were achieved and Iraq has since been living under severe U.N. sanctions. U.S. killed in action numbered 148, 28 Americans were killed by a Scud missile hit on a barracks in a rear area. U.S. wounded numbered 213. The other Coalition countries reported a total of 76 killed in action and 126 wounded. Iraqi casualties are estimated to have been at least 80,000 troops killed (and some civilians) but Iraq released no official numbers of its killed or wounded.

Following the war, the U.S. Army Field Artillery School published observations on various aspects of artillery performance in Desert Storm, which showed that in its first combat test, AirLand Battle doctrine was thoroughly valid. It confirmed for example, that massed fires had overwhelmed the enemy in Iraq. Target acquisition systems and command, control and communications systems all worked in an integrated fashion to silence the Iraqi artillery and armor and pave the way for Coalition infantry and armor successes against a battered enemy. One Iraqi artillery commander noted he lost ten percent of his guns to air attack and the remaining 90 percent to massed fires of ground artillery and rockets. In one instance, Bravo Battery, 3rd Battalion, 82nd Field Artillery, fired four Copperhead rounds that were laser-guided by an OH-58D helicopter, which destroyed two Iraqi tanks and a tracked antiaircraft weapon system. Army artillery units fired more than 57,000 rounds during the 100-hour war. The M109 self-propelled 155-mm howitzers accounted for more than 43,000 (75 percent) of the total. Other reports indicated MLRS fires were devastating and exceeded expectations when used in conjunction with Q-36 and Q-37 Firefinder radars.

One generally unexpected hazard after the end of hostilities, was the so-called "dirty battlefield." This condition was the result of using munitions containing bomblets or submunitions released but not detonated on impact. These munitions included dual-purpose improved conventional munitions (DPICM), MLRS submunitions and Air Force cluster bombs. Some target areas contained dud bomblets on the ground, which exploded on contact by vehicles or personnel, causing minor damage and some injuries.

There were other problems, also. Iraq fielded several cannons that outranged U.S. cannons by a considerable margin. The Iraqi GHN-45 and G-5 howitzers had a range of 39.6 kilometers (24.6 miles) compared the M109's rocket-assisted projectile range of 23.5 kilometers (14.6 miles).

There was no doctrine for executing fire support at the field army and joint (Army/Air Force) level. It would have to be written. Combat service teams were shifted among artillery units due to the rapid infantry and armor advances in the Gulf War. The teams often could not adequately perform their service functions for any unit to which they were attached. MLRSs were overworked; everyone wanted them — corps, brigade and division. Most commanders saw the MLRS as the weapon of choice in counterfire work up to about an 18-mile range, although it was not accurate enough for close fires. It had been the same story since Revolutionary times — all units wanted more artillery firepower. There were also some ammunition resupply problems. Ammunition delivery often could not keep up with the rapid rate of firing by the artillery. There were also communications problems. The U.S. forces generally lacked a reliable means of secure, long-range radio communications for the highly mobile operations of Desert Storm. Field artillery units often found themselves beyond FM radio range, and the AM radios were not only in short supply, but also were unreliable and did not support digital communications.

Some field artillery vehicles were shown to be substandard in performance. Command posts in five-ton expandable vans were unable to keep up with the tactical command post of the maneuver forces. The Fire Support Vehicle (M981) also was unable to keep pace with the Bradley fighting vehicles and Abrams tanks, and the M981 had turret maintenance problems. The ammunition carrier vehicle (M548) was often the slowest vehicle in the force and, when fully loaded, determined the maximum speed of movement.

The new ATACMS, its projectiles fired by MLRS, was used against high-value targets (Scud missile sites, ground-to-air missile sites, logistical bases, important bridges) with great success, destroying every identified target. Artillery modernization efforts of the 1980s paid dividends in this mid-intensity conflict in the Persian Gulf. The Firefinder radar systems worked well, but also had some problems. The Q-37 had mobility problems and none could complete the cross-desert attack into Iraq with the XVIII Corps. The Q-36 bore the brunt of the workload and performed well. The Q-36, however, was overly sensitive and detected both friendly and enemy targets so careful screening was always necessary.

C Battery, 3rd Battalion, 82nd Field Artillery, during Desert Shield. (Photo courtesy of Kenneth R. Knight)

First Lieutenant Al Kearse "Laying the platoon" during Desert Storm. (Photo courtesy of Kenneth R. Knight)

Artillery preparation fired by the 82nd Artillery in support of the 1st Cavalry Division. (Photo courtesy of Kenneth R. Knight)

National Guard Artillery

The National Guard is one of the reserve organizations of the U.S. Army. An Air National Guard was established in 1947 with the separation of the Air Force from the Army. National Guard origins are from the pre-Revolutionary War Colonial militias, and its organization today is similar to that of the U.S. Army. The first organized artillery unit appears to be the "Ancient and Honorable Artillery Company," organized in 1638 by the Massachusetts Bay militia. By the middle 1700s, various Colonial cities (Boston, Massachusetts; New York, New York; Philadelphia, Pennsylvania; Newport, Rhode Island and Charleston, South Carolina) had organized volunteer artillery companies separate from the militia infantry. Some men in these artillery units had been trained by British regular artillerymen in the use of their cannon.

Revolutionary War

The first regiment of artillery was organized in the spring of 1775, from Massachusetts and Rhode Island militias, and became the Continental Artillery Regiment, the first in the new Continental Army. Other rebellious colonies soon followed suit. During the war, Colonial militia artillery batteries fired side by side with Continental units in the defense of or attack on various cities (Charleston, South Carolina; Savannah, Georgia; Philadelphia, Pennsylvania; Trenton and Princeton, New Jersey). Henry Knox, self-taught, was Washington's Chief of Artillery during the Revolution, and saw the artillery through to the end of the war.

Following the Revolution, the Army artillery was reduced to three batteries, and the militiamen became civilians again. By the Militia Act of 1792, state militias were required to organize one artillery battery for each division of infantry. Their cannons sometimes came from the regular Army, sometimes issued by the state. Militiamen were unpaid. Militias were organized into battalions and regiments over the years, but not until 1798 did the regular Army organize the 1st and 2nd Artillery Regiments.

War of 1812

The entire militia was mobilized for service in this war, and most artillery units were positioned primarily in the east, where they could register on landing British troops. The Army grew from less than 12,000 men to 33,000 men by war's end. Artillery slowed the British attack on Washington, but could not stop them from taking and burning the city. It did blunt the failed British attack on Baltimore, Maryland, however. In the west, artillery often played a significant, even decisive role in some of the battles with the British, but was never totally effective because of the rugged forest. The exception perhaps, was at New Orleans, a battle fought after the war had ended. Here, artillery was the deciding factor in the American victory. Direct, observed fire, of course, was the rule in these days. The war spurred the growth of the numbers of militia units in the young United States. Over the years, the old militia (whose men were required by law to serve) was replaced by all-volunteer units that trained hard and took great pride in their unit. They developed functions before the Civil War, some of which characterize the Guard today: ceremonial parades, manpower to help in disaster relief and a force available against riots, for example. Close-order drill and shooting accuracy competitions were favorite exercises among units in those days of peace.

Civil War

The organized militia in the north, including artillery units, entered Federal Service with the outbreak of war in April 1861. Two hundred twenty-two batteries were organized into 41 artillery regiments and all entered Federal Service. Southern militia units were also the beginnings of the Confederate armies. Many of today's Field Artillery regiments had their origins at this time. The 101st Field Artillery Regiment came from the old Boston Light Artillery of Civil War times; the 141st Field Artillery of today was the old Washington Artillery of New Orleans, which saw action in Lee's Army of Northern Virginia, for example. Today, 15 regiments of National Guard field artillery carry Civil War battle streamers.

Following the war, these organized militias became known as the National Guard and continued their drills and competitions, now supplied with excess guns, limbers and caissons and other artillery equipment from the Civil War. Guard artillery was not called up to serve in the Spanish-American War, although batteries from California and Utah were sent to the Philippines to combat and suppress the Insurrection by Philippine rebels seeking independence following the War. These units received invaluable guerrilla-style combat experience in an inhospitable, tropical climate. Other Pennsylvania batteries saw service in the Puerto Rico campaign during this war. By the Dick Act of 1903, the National Guard became independent of the regular Army and was subject to mobilization only by order of the state governors. The Guard artillery expanded so that by 1916, 108 field artillery batteries had been organized, although twice that number had been authorized. Also, under the National Defense Act of 1916, the Guard became subject to Federal mobilization and later amendments made the Guard a Reserve Component of the U.S. Army. All artillery units (9,500 men) were mobilized in 1916 for service during the Mexican border conflict, and they trained in the southwestern deserts, but without entering combat. All units were home early in 1917 in time for America's entry into World War I in April.

World War I

With America in the Great War, National Guard artillery quickly expanded to 50 regiments numbered from 101 to 151. These supported 17 Guard infantry divisions numbered 26 through 42. Up to this time, Guard artillery regiments were identified by names, such as the 1st Massachusetts Field Artillery Regiment, but now Guard regiments were numbered such that the Massachusetts regiment became the 101st U.S. Field Artillery Regiment. The Guard units were issued French 75s and began intensive training. The Guard infantry divisions were sent to France early in the war, and Guard artillery provided about 40 percent of the U.S. artillery regiments that served in France. One of their biggest battles was the Meuse-Argonne campaign in the fall of 1918, where Guard artillery supported the massive attack that rolled back the German armies to eventually end the war. One of these regiments was the 121st Field Artillery Regiment (Wisconsin) with 155-mm howitzers, which distinguished itself in this and five other major campaigns in France, primarily with the 32nd Infantry Division. Its battalions were redesignated in 1942, and fought with distinction in both Europe and the Pacific in World War II.

Following the Great War, National Guard field artillery was reduced to fewer than 20 under-strength regiments, nationwide. Through the 1920s and 1930s lack of funds saw deterioration in Guard equipment and training, although the Army did convert, by 1940, almost all Guard artillery batteries to motorized (truck towed) units. The National Defense Acts of 1920 and 1933 further extended federal control over the National Guard so that the President was given authority to order the National Guard to duty without going through the State Governors. With a war raging in Europe, the National Guard was called up in September 1940.

World War II

During World War II, the Guard supplied 156 battalions of artillery in 78 regiments; 72 battalions were divisional artillery. The first artillerymen in the Pacific Theater were Guardsmen. The 147th Field Artillery was at sea and diverted to Australia after the Japanese attack on Hawaii. The 2nd Battalion, 131st Field Artillery, of Texas, was in Java in March 1942. The 32nd Division Artillery and the 41st Division Artillery fought in New Guinea in late 1942, and the 221st Field Artillery Battalion (Massachusetts) was on Guadalcanal in January 1943. In the Pacific War with Japan, Guard artillery units were in every major U.S. Army battle from the Aleutian Islands to the Philippines, to Okinawa. The 175th Field Artillery Battalion (Minnesota) fought with the 34th Division in Algeria in November 1942, and Guard units fought in every major battle in the Mediterranean and Europe including Italy, France, Belgium and finally Germany. Reid W. McNary was a First Sergeant with the 981st Field Artillery Battalion, First Army (originally 144th Field Artillery), California Army National Guard, and he recalls:

In early June [1944] we were loaded aboard for the most awesome scene I had ever seen; ships, boats, planes and then the beaches. We landed at Utah Beach [on June 14], into St. Mere Eglise, then joined the battle for Cherbourg, and our weapons never failed us, and we were successful in the capture of the port. We received special commendations from General Bernard Montgomery for our efforts.

Our Long Toms [155-mm gun] continued across France, into Belgium, The Bulge and into Germany. We were on the Elbe river, met the Russians. Those Long Toms never let us down. It was said that it was one shell over, one under, and then the next was in your vest pocket. We captured a German Doctor [and] he stated that in five years of combat he had never endured such devastating fire.

.....we were stationed in the Rhur Valley, and as the forward observer for my Battalion, I had the most wonderful view of the Rhur Dams. They were every artilleryman's dream of a target. I watched those dams, and the Colonel said, "Sergeant, if I see those dams shelled, you will spend a long time in the stockade, because I will know who shelled them." But damn, what a beautiful target!

William D. Smith was in the 981st also; he recalls his training on the 155-mm at Fort Lewis, Washington:

Leaving New Cumberland we took off by train for Fort Lewis, Washington. It took two days for us to arrive. Here we became part of the 981st Long Tom 155-mm Heavy Field Artillery; this unit was originally the 144th Field Artillery of California.

We had three months of training here to learn all about the guns as to how to put them in position. We did a lot of firing, and I set the quadrant most of the time. The shells weighed 96 pounds and powder charges weighed 33 pounds. Two men carried the shell on a tray, and it was placed in the gun with a large cone-shaped ramrod. The powder was then placed in the gun, and the breach of the gun was closed. When the proper elevation was set, the quadrant man would call out "Fire" [and] the shell was on its way. Short- or long-range fire, we were very accurate up to 12 miles.

Korean War

Between 1951 and 1953, a total of 20 Guard artillery battalions saw combat in Korea although many more entered Federal Service. Thousands of replacements were sent to Korea to relieve the original contingents and Guardsmen supported both Army and Marine infantry units throughout the war. Guard units were reduced following the Korean War, but 17 battalions and four headquarters batteries were called into Federal Service during the Berlin crisis of 1961, when the Soviet Union shut down Western Bloc road traffic to Berlin across Communist East Germany.

Vietnam War

Only five Guard artillery battalions were ordered to active service in Vietnam: 1st Battalion, 487th Artillery (Hawaii); 2nd Battalion, 138th Artillery (Kentucky); 1st Battalion, 211th Artillery (Massachusetts); 3rd Battalion, 197th Artillery (New Hampshire); and 2nd Battalion, 130th Artillery (Kansas). The Kansas and New Hampshire battalions were in Vietnam in the fall of 1968 and stayed for one year. Other Guardsmen were ordered to Vietnam as replacements.

Following the Vietnam War, the National Guard has been called to duty several times during the years of relative peace, at times for riot control. Part of the Arkansas 142nd Field Artillery Brigade was called to Federal Service to quell riots by Cuban refugees at Fort Chaffee, Arkansas, in June 1980. The 40th Infantry Division Artillery was ordered to Federal Service in May 1992, to put down rioting in Los Angeles. Through the 1980s, there were organized a total of 93 battalions (39 divisional and 54 corps) plus 18 brigades and one headquarters battery.

Persian Gulf War

In the fall of 1990 during Desert Shield, the buildup of forces in Saudi Arabia to oppose Iraq's incursion into Kuwait, two National Guard field artillery brigades were ordered to Federal Service and sent to Saudi Arabia.

(1) Headquarters and Headquarters Battery, 196th Field Artillery Brigade (Tennessee), composed of 1st Battalion, 201st Field Artillery, 155-mm self-propelled howitzer (West Virginia); 1st Battalion, 623rd Field Artillery, 8-inch self-propelled howitzer (Kentucky); and 1st Battalion, 181st Field Artillery, 8-inch self-propelled howitzer (Tennessee).

Kentucky National Guard overall view of Fire Base Denise, manned by members of Battery C, 2nd Battalion, 138th Artillery, in Vietnam, January 1969. (Photo courtesy of the National Archives)

A 155-mm self-propelled howitzer of the 1st Battalion, 15th Field Artillery, crossing the Imjin River near Chon Gok, Korea, in March 1995. (Photo courtesy of Darrell J. Weatherford)

(2) Headquarters and Headquarters Battery, 142nd Field Artillery Brigade (Arkansas), composed of 1st and 2nd Battalions, 142nd Field Artillery, 8-inch self-propelled howitzers (both battalions from Arkansas); and 1st Battalion, 158th Field Artillery, Multiple Launch Rocket System (Oklahoma). The 142nd Brigade supported the 1st Armored Division (British) and was given extremely high marks for its performance in combat. Three other battalions were mobilized, but remained in the U.S. These were: 1st Battalion, 141st Field Artillery (Louisiana); 1st Battalion, 230th Field Artillery (Georgia); and 2nd Battalion, 114th Field Artillery (Mississippi).

The following paragraphs are from Lieutenant Colonel Leonid Kondratiuk's excellent pamphlet on the history of the National Guard Field Artillery:

Currently, the Army National Guard field artillery force consists of 32 divisional battalions, 19 assigned to separate infantry and armored brigades, one roundout battalion, 36 assigned to 16 field artillery brigades for a total of 88 battalions. The Guard's field artillery structure will grow during the next year (by 1998) to 24 divisional battalions, seven target acquisition batteries, eight divisional 155-mm batteries, 14 battalions assigned to enhanced brigades, two assigned to theater defense brigades, one roundout battalion and 49 battalions assigned to 17 field artillery brigades for a total of 90 battalions. The National Guard field artillery force will be the largest in the Army and will consist of 65 percent of the force.

For over 368 years, the National Guard has provided field artillery in both Colonial and American wars. It has traditionally maintained most of the Army's field artillery force during peacetime and has been the largest source for personnel and units at the beginning of our nation's wars. Because of the interchange of personnel and training, the relationship between Active and National Guard artillerymen is close and one of mutual respect. It has been this way since 1775. The National Guard will continue its historic mission of maintaining the largest part of the Army's field artillery force. The tradition continues.

Field Artillery Stories

The Legend of Saint Barbara

According to legend, our patron saint was the beautiful daughter of Dioscorus, a nobleman of the Roman Empire, believed to have lived in Nicomedia in Asia Minor in the Third or Fourth century, A.D. To limit Barbara's exposure to Christianity and encourage her development as a zealous pagan, her father kept her shut up in a tower, lighted by only two windows. But from these windows she looked out upon the surrounding countryside and marveled at the living things. She concluded they all must be part of a master plan and the idols of wood and stone worshipped by her parents had to be condemned as false. She obtained instruction in Christianity and was baptized. In token of her faith while her father was away, she had another window pierced in the tower, making three, symbolizing the Holy Trinity. On his return, Dioscorus asked why she had made this change, and Barbara acknowledged her conversion. Despite his threats, she refused to renounce Christianity. He delivered her to a Roman magistrate, who even with torture, also failed to persuade her. Dioscorus, himself, then took his daughter to a high mountain, where he beheaded her. Afterward, as he descended the mountain, he was caught in a sudden violent storm. In a blinding flash he was consumed by lightning.

As a logical consequence, Barbara came to be regarded as the sainted patroness of those in danger from thunderstorms, fire, explosions — that is to say, sudden death. Given the questionable reliability of early cannon — misfires, muzzle bursts and exploding weapons were not uncommon — it's easy to see why our predecessors sought the protection of Saint Barbara. She has protected us well ever since.

The Story of Molly Pitcher

An artillery wife, Mary Hays McCauly (better known as Molly Pitcher) shared the rigors of Valley Forge with her husband, William Hays. Her actions during the battle of Monmouth (June 28, 1778) became legendary. That day at Monmouth was as hot as Valley Forge was cold. Someone had to cool the hot guns and bathe parched throats with water.

Across that bullet-swept ground, a striped skirt fluttered. Mary Hays McCauly was earning her nickname "Molly Pitcher" by bringing pitcher after pitcher of cool spring water to the exhausted and thirsty men. She also tended to the wounded and once, heaving a crippled Continental soldier up on her strong young back, carried him out of reach of hard-charging Britishers. On her next trip with water, she found her artilleryman husband back with the guns again, replacing a casualty. While she watched, Hays fell wounded. The piece, its crew too depleted to serve it, was about to be withdrawn. Without hesitation, Molly stepped forward and took the rammer staff from her fallen husband's hands. For the second time on an American battlefield, a woman manned a gun. (The first was Margaret Corbin during the defense of Fort Washington in 1776.) Resolutely, she stayed at her post in the face of heavy enemy fire, ably acting as a matross (gunner).

For her heroic role, General Washington, himself, issued her a warrant as a noncommissioned officer. Thereafter, she was widely hailed as "Sergeant Molly." A flagstaff and cannon stand at her gravesite at Carlisle, Pennsylvania. A sculpture on the battle monument commemorates her courageous deed.

Fiddler's Green

Imagine, if you will, a starry night in southwestern Oklahoma just after the Civil War. Nestled in the shadows of the Wichita Mountains is a battery of smoothbore cannon camped for the night. As the campfires dim and the flasks of rum and lemon empty, the conversation turns to life in the hereafter. A rugged, old chief of section is surprised to learn that all present have not heard of the special destiny of Redlegs. As the young cannoneers listen intently, he shares with them the legend of Fiddler's Green.

The chief of section explains that the souls of the departed eventually end up in heaven or hell. Heaven lies about six miles down the dusty road to eternity, and Redlegs get there by turning left at the first crossroad. From this same junction, hell is about eight or nine miles straight ahead. The road's easy to identify: it's the one paved with good intentions. A little way down the road to hell, there is a sign pointing to a trail that runs off to the right of the main road. It reads "Fiddler's Green — Artillerymen Only."

He then teaches them the following poem:

Halfway down the trail to hell,
In a shady meadow green,
Are the souls of many departed Redlegs
Camped near a good old-time canteen.
And this eternal resting place
Is known as Fiddler's Green.

Though others must go down the trail
To seek a warmer scene,
No Redleg ever goes to hell,
Ere he's emptied his canteen.
And so returns to drink again,
With friends at Fiddler's Green.

The campfires die out, and the Redlegs doze off to sleep, knowing Fiddler's Green awaits them and all their cannon-cocking brethren in the life hereafter.

Field Artillery Officer Candidate School

With an increasing demand for artillery officers at the outbreak of World War II, the Field Artillery Officer Candidate School was officially opened in 1941. The first class of Field Artillery officer candidates reported to Fort Sill, Oklahoma, on July 10, of that year. It was billeted in tents in the vicinity of the present 95th AG.

The school was established to produce junior officers who could successfully lead artillery elements in combat. Early precedents concerning officer candidates and their treatment were set. General R.N. Danford, World War II Chief of Artillery, directed the candidates be worked as hard as possible to weed out those who could not stand the pressure.

Artillerymen firing their field pieces. (Photo courtesy of Fred C. Battles)

The school offered a 13-week course for warrant officers and enlisted men desiring to become second lieutenants. Captain Carl H. Jark was assigned as the school's first commandant. Five years later, on December 12, 1946, with the graduation of Class 179, the school closed, having commissioned more than 26,000 second lieutenants.

The Korean mobilization resulted in the OCS' being reopened on February 21, 1951, when the first class since World War II reported to attend the new 23-week course. Fifty-three candidates started with the class and 28 were commissioned. On July 7, 1973, after the American withdrawal from the War in Vietnam, Fort Sill's Officer Candidate School officially closed with the graduation of 26 Field Artillerymen. During the 32-year history of Artillery OCS at Fort Sill, 47,500 second lieutenants graduated.

The Artillery Officer Candidate School was renamed the Artillery and Guided Missile Officer Candidate School on May 4, 1955, and further changed in July 1967 to the U.S. Army Artillery and Missile Officer Candidate School.

During its history, the Field Artillery OCS has produced a remarkable number of heroes. Two of these men warrant special mention. First Lieutenant James E. Robinson, Jr., a 1943 graduate, was the only artillery officer to be awarded the Medal of Honor during World War II. His award was presented posthumously for his actions in an attack near Untergresheim, Germany, in 1945. The old OCS area, where the 95th AG is located today, was named "Robinson Barracks" in his honor, on April 15, 1953. Second Lieutenant Harold B. Durham, Jr., was awarded the Medal of Honor for his actions while serving as a forward observer in Vietnam in 1967.

A total of 130 U.S. Army artillerymen have won Medals of Honor in all wars from the Civil War to the present. Ten Marines and 14 Navy men (fighting as artillerymen on land) have won the Medal. One hundred two of these 154 were enlisted men at the time of winning the Medal.

United States Field Artillery Association

Rather than paraphrase the work of another author it seems appropriate to simply restate the "Brief History of the United States Field Artillery Association" as it appeared in the elegant annual brochure of the Association (pp. 15-16). This publication also included chapters on the Aims, Challenges, Programs and Special Projects of the Association in addition to much other Association information.

The Early Days

The idea for an association that would unite the gunners of the newly created Field Artillery Branch resulted from the discussions of Colonel Montgomery M. Macomb, Commander of the 6th Field Artillery, and his Adjutant, Captain William J. Snow. As these two distinguished artillerymen took their early morning rides at Fort Riley, Kansas, they spoke of an organization that would promote the field artillery and enhance the professionalism of artillerymen. Fortunately, these men were not just idle dreamers; they made things happen. Early in the spring of 1907, Captain Snow set out to make the association a reality. His first step was to dispatch a circular letter to the 180 regular officers of the field artillery, seeking their reaction to his proposal, which included as its principal feature the publication of a branch journal. Although a number of respondents were concerned about financing such a magazine, most officers heartily approved the idea. Captain Snow proceeded and, after a lengthy exchange of correspondence with the War Department, obtained appropriate approval.

Early in 1909, Captain Snow joined a committee including Major John E. McMahon and Captain W.S. McNair to draft a constitution. In a slightly modified form, the constitution was introduced to the Regular Army and National Guard Field Artillery officers attending the summer "Encampment and Maneuvers" at Fort Riley, during the following year. On June 7, 1910, these 86 artillerymen became the charter members of the Association established under the provisions of a constitution that sought:

- *To promote the efficiency of the Field Artillery by maintaining its best traditions.*
- *To cultivate with the other arms a common understanding of the powers and limitations of each.*
- *To foster a feeling of interdependence among the different arms and of hearty cooperation by all.*
- *To promote understanding between the regular and militia forces by a closer bond.*

The charter members elected the first Executive Council, which included such luminaries as Colonel Macomb and Captain Fox Conner. Unlike other branch associations, the new Field Artillery Association involved men from both the Regular Army and the National Guard. This tradition of corporateness continues today.

In November 1910, the Executive Council elected Captain Snow as its Secretary-Editor. With the assistance of his "enthusiastic and indefatigable" clerk, Mr. Charles S. West, Captain Snow soon produced the first issue of the Field Artillery Journal *— January-March 1911.*

Colonel Macomb's and Captain Snow's dream had become a reality, and that reality nurtured the growing branch General Snow would later lead as the first Chief of Field Artillery.

Evolution and Transformation

Between January 1911 and June 1950, the Field Artillery Journal, *the most conspicuous product of the Field Artillery Association, had 19 different part-time or full-time editors ranging in rank from Captain to Colonel. The frequency of the* Journal*'s publication varied, its finances were often precarious and the magazine's circulation fluctuated dramatically according to the times. But the Association, through its* Journal, *always provided its eager readers with timely, accurate information as well as thought-provoking pieces, including vanguard Colonel Ardant du Picq, General Frederick Georges Herr, Colonel Conrad H. Lanza and Major Fairfax Downey.*

Shortly after the close of World War II, the government announced withdrawal of its support of professional associations. The financially plagued Association faced a crisis. Without a professional military staff, the Journal *and the Association simply could not make ends meet. So in late 1949, the Executive Council recommended to the membership that the Association "merge with other associations to form a single combat arms association that would produce a combined monthly journal." At a special meeting held at the Army-Navy Club in Washington, D.C., on April 3, 1950, the Association's Executive Council announced that more than 97 percent of the members canvassed favored a merger with the Infantry Association to form the Association of the United States Army (AUSA). With the publication of the May-June 1950* Journal, *the final issue of an uninterrupted 40-year series, the Field Artillery Association took on a new, but disguised image for a period of 24 years.*

The Renaissance

Although transformed, the Field Artillery Association and its Journal *did not die. The spirit of camaraderie and the need for such a professional organization and intellectual forum remained a smoldering fire. On November 17, 1974, the 199th anniversary of the formation of the United States Field Artillery, this fire sprang to life as a rejuvenated Field Artillery Historical Association, which distributed a new* Field Artillery Journal *bimonthly publication. In 1980, Major General Jack N. Merritt, the Commanding General of Fort Sill, stoked the fire as he championed the cause of a separate, revitalized Field Artillery Association. This independent organization, called the United States Field Artillery Association, held its first general membership meeting on October 21, and adopted the same objectives as those first set forward in 1910, by then Captain William J. Snow.*

In May 1986, the Army Publications Review Committee recommended the Journal *for elimination with 40 other publications. But the Commanding General of the Training and Doctrine Command (TRADOC) decided to continue the publication in an economical professional bulletin format. The July-August 1987 issue was the first to comply with the TRADOC regulations, including changing the title to* Field Artillery *with the sub-title* A Professional Bulletin for Redlegs, *the adoption of the Army's professional bulletin numbering system and the elimination of any purely editorial or human interest articles.*

With the March-April 1996 edition, the Association began publishing the FA Journal. *The Association staff now adds the content of the "Forward Observer" to* Field Artillery *to publish a new and upgraded* FA Journal. *This forum gives the Association the space it needs for editorial comment, news items and announcements appropriate for its mission.*

Unlike its predecessor, the new Association has not confined itself to the publication of Field Artillery. *Today, the Association is executing many programs that not only promote the field artillery in general, but also foster the development and recognition of members of every component of the field artillery community.*

The Association's local chapters that dot the globe provide gunners and their associates a forum for the debate of contemporary issues as well as for the veneration of Redleg traditions; the Association's Orders of Saint Barbara and the Artillery Order of Molly Pitcher recognize the professional accomplishments of those who serve the field artillery and their spouses. The growing Association involvement with retired gunners and members of defense industry ensures that today's leaders of the field artillery preserve our past, explain our present and assure the future.

Since its humble beginnings on horseback at Fort Riley in the first decade of this century, the United States Field Artillery Association has served important purposes and contributed to the development of the world's best Field Artillery. Today's Association continues this proud Redleg tradition.

Statement of Purpose:

The purposes of the United States Field Artillery Association are:

- To promote the field artillery.
- To disseminate knowledge of military art and sciences, with special attention to the role of field artillery and fire support in warfare.
- To promote the professional development of its members.
- To foster and preserve the spirit, traditions and solidarity of the field artillery.

To achieve these purposes, the Association has developed several programs:

1) Individual Membership Program.
2) Corporate Membership Program.
3) Local Chapter Program.
4) Professional Recognition Program.

Individual membership is open to all military personnel, officer and enlisted, active or retired, service academy cadets, students in other military schools and to civilians who subscribe to the Association's purposes.

Corporate membership is open to all industries that participate in development of components of fire support systems or who contribute to the improvement of the field artillery by providing concepts, studies and services.

Local chapters are the primary means by which the Association seeks to achieve its purposes. They provide a forum for the development of future generations of Redlegs and a means for spreading the field artillery's story. They also provide a convenient, well-supported organization for local field artillery leaders to use to preserve and expand Redleg traditions — build teams. The Association's latest publication lists 68 local chapters.

The Professional Recognition program provides recognition of members of the field artillery community who have, through their conspicuous service and achievement, advanced the purpose and objectives of the United States Field Artillery Association. Any member of this community may be nominated to receive appropriate Association-sponsored awards. The awards are:

- The Ancient Order of Saint Barbara recognizes those individuals who have performed long-term, exceptionally conspicuous service for or on behalf of the field artillery and whose careers embody the spirit, dignity and sense of sacrifice and commitment associated with Saint Barbara.
- The Honorable Order of Saint Barbara recognizes those individuals who have demonstrated the highest standards of integrity and moral character; displayed an outstanding degree of professional competence; served the United States Army or Marine Corps field artillery with selflessness; and contributed to the promotion of the field artillery in ways that

A recreation of Lamb's artillery of 1775. The 3-pounder Verbrugen field piece was made from an original taken at Saratoga. (Photo courtesy of John R. Vilven)

stand out in the eyes of the recipient's seniors, subordinates and peers alike.

• The Artillery Order of Molly Pitcher recognizes individuals whose voluntary service has improved the field artillery community and whose performance is in keeping with the spirit of sacrifice and devotion exemplified by Molly Pitcher.

In addition to these activities, the Association sponsors certain special projects, among them:

• Field Artillery Luncheons at the Fort Myer Officers' Club, which attract members in the Washington, D.C., area. The Chief of Field Artillery is the featured speaker who reviews the branch's priority programs.

• The Distinguished Graduate Award is presented to specific Field Artillery School course graduates and to the top-ranking ROTC and West Point cadets who select field artillery as their branch of service.

• The History Writing Contest started in 1985 as an annual event. The contest is open to anyone interested in the history of the King of Battle and is judged on relevance to field artillery tactics, usefulness to today's artillerymen, historical accuracy, writing effectiveness and originality.

• The *FA Journal* is the Association's major publication. Beginning with the March-April 1996 edition, two former publications, "Forward Observer" and *Field Artillery* were combined to produce the new *FA Journal*. This enhanced version is published in full color, and includes editorial comments, news items and announcements, advertising, chapter news, reunion notices and Association-related articles.

• Historical Paintings Series is a group of eight paintings commissioned by the late Mr. John J. McMahon, 1923-1992, long-time member and benefactor of the United States Field Artillery Association. The paintings depict scenes from the Revolutionary War, Marines at Veracruz, Civil War, World War I, World War II, Korean War, Vietnam War and Desert Storm of the Persian Gulf War. The Association has printed lithographic copies of these paintings for sale.

References

In any work of prescribed length such as this one, only some of the information in the many references could be incorporated and that is always a judgment call by the author. In addition to several general historical sources, the two most used artillery-specific references are the works of Boyd L. Dastrup listed below:

Dastrup, Boyd L., *King of Battle*. Office of the Command Historian, United States Army Training and Doctrine Command, Fort Monroe, Virginia. 1992.

Dastrup, Boyd L., *Modernizing the King of Battle, 1973-1991*. Office of the Command Historian, United States Army Field Artillery Center and School, Fort Sill, Oklahoma. 1994.

For information on Field Artillery in the Persian Gulf War, the October, 1991, issue of *Field Artillery,* Special Desert Storm Edition, was very helpful.

Many more references were used (sent to the author via the publisher) from many artillerymen. Their contributions are gratefully acknowledged. Among these contributors and their subjects (the list is not all-inclusive, nor in any particular order) are:

United States Field Artillery Association; Eugene L. O'Brien (155-mm Artillery); Lieutenant Colonel David T. Zabecki (*American Artillery and the Medal of Honor*); Major W.D. Crittenberger (420th Field Artillery Battalion); William D. Smith (981st Field Artillery Battalion); Reid W. McNary (981st Field Artillery Battalion); John Stafford (1st Battalion, 121st Field Artillery Regiment); George J. Ellis (780th Field Artillery Battalion); Robert C. Baldridge (*Atomic Annie*); Colonel Griffin N. Dodge (11th Air Assault Division); Major Robert B. Leach (1st Battalion, 5th Field Artillery); Colonel Kenneth R. Knight (82nd Field Artillery); Command Sergeant Major William Perry (Operation Just Cause); Lieutenant Colonel Leonid Kondratiuk (National Guard); Colonel Melvin H. Rosen (Philippine Scouts).

Artillery crew of the 46th Field Artillery Battalion take up various positions at a forward observation post at Westdown Fields, Tidworth Wilts, England, October 2, 1943. (Photo courtesy of the National Archives)

Men of the 39th Field Artillery, 3rd Infantry Division, dug in on the beach at Hungnam, Korea, fire at Chinese Communists, December 23, 1950. (Photo courtesy of the National Archives)

Field Artillery Association Roster

Aaltonen, Tauno O.
Aaron, Daniel G.
Aaron, Larry D.
Aas, Robert K.
Abel, Stephen G.
Abernathy, William L.
Abuzeide, Philip L.
Achs, Jack H.
Acosta, Thomas W. Jr.
Acton, Jerry R.
Adair, Charles W.
Adair, Robert B.
Adam, Klaus M.
Adams, Kevin J.
Adams, Lewis D. Jr.
Adams, Mark E.
Adams, Michael C.
Adams, Paul C.
Adamson, David J.
Addison, Richard L.
Addy, Joseph M.
Adelberg, Michael A.
Adkins, Brian E.
Adler, Nicholas S.
Agans, Robert C.
Agee, Gregory O.
Agron, Gary A.
Ahrens, David A.
Airey, James M.
Aitken, David M.
Aitken, Tristan N.
Akechi, Masaru
Akers, Donald K.
Akin, Thomas S.
Akselrud, Michael
Albaneze, Michael A.
Alch, Steven T.
Alcock, Edward J. Sr.
Alday, Clint L.
Alderfer, David W.
Alexander, Lawrence R.
Alexander, Steven R.
Alexander, Timothy H.
Alexander, Willie A.
Alfaro, Armando J.
Alger, Don R.
Alhouse, Robert D.
Alicea, Randell
Alkire, Eugene S.
Allard, Chris L.
Allen, Charles B.
Allen, Charles E.
Allen, Edward D.
Allen, George A.
Allen, Jason S.
Allen, Robert L.
Allen, Robin R.
Allen, Stephen P.
Allen, Timothy J.
Allen, William C. III
Allison, Clay F.
Allwine, David L.
Almquist, Martin S.
Alonso, Joseph
Altersitz, Larry A.
Altman, S. M.
Amber, Samuel H.
Amoros, Edward G.
Anast, Bruce J.
Anderson, Alan J.
Anderson, Darran T.
Anderson, David
Anderson, David S.
Anderson, Edward G.
Anderson, Edward G. IV
Anderson, John K.
Anderson, Mark E.
Anderson, Mark T.
Anderson, Matthew R.
Anderson, Phillip R.
Anderson, Randall J.
Anderson, Robert C.
Anderson, Rodney O.
Anderson, R. Reed
Anderson, Sid D.
Anderson, William M.
Andrew, Robert B.
Andrews, Curtis W.
Andrews, Eric
Andrews, Gary W.
Andrews, John M.
Andrews, Phillip C.
Andrews, Ronald J.
Angel, George W.
Angell, Michael V.
Anger, Allan W.
Anglin, Darren W.
Annen, David M.
Antis, Robert M.
Antone, Jeffrey
Antonietti, Patrick M.
Anzalone, John E.
Apgar, William I.
Appleget, Jeffrey A.
Arblaster, Steve R. II
Archambault, Bruce A. Jr.
Archbold, Cecilio M.
Archer, David W.
Arguello, Evan R.
Armour, David T.
Armstrong, Charles E.
Armstrong, Gerald T.
Armstrong, Paul C.
Armsworth, Todd A.
Arnold, Brian R.
Arnold, Kevin J.
Arnone, Robert F.
Arntz, Stephen J.
Aronson, Richard
Arsenault, Michael R.
Artigliere, William M.
Arvidson, Mark D.
Arvizu, Ismael
Ash, Michael D.
Ashley, Michael P.
Ashmore, William C.
Ashworth, Eric L.
Aswell, Paul L.
Atchison, James W.
Atkinson, Lorenzo D.
Audens, Michael W.
Aufill, John S.
Aust, Henry J.
Austin, Douglas W.
Avance, Jimmie D.
Averill, Allen A.
Aversano, Mike G.
Ayala, Milton R.
Babb, Graham
Babcanec, Andrew S.
Babers, Donald M.
Bach, Steven E.
Bachar, Jay B.
Bachmurski, Stanley M.
Bacon, Edward S.
Bacon, James D. H.
Bacon, Peter K.
Badami, Victor
Badger, Ralph E.
Bado, Christopher M.
Bagby, Byron S.
Baham, Gaylon T.
Bailes, Robert L.
Bailey, Charles R.
Bailey, Edward A.
Bailey, Jerrold L.

Bailey, Ronald M.
Bailey, Stephen L.
Bailey, Stewart L.
Baird, Brian
Baird, John
Baird, John S.
Baird, Stephen W.
Baker, Christopher D.
Baker, Edward E.
Baker, Mary (Annie)
Baker, Peter R.
Baker, Richard J.
Baker, Timothy J.
Baldridge, Robert C.
Balestrieri, Ralph R.
Ball, Charles T.
Ball, Chester E.
Ball, Robert D.
Ball, Shawn D.
Balogh, William
Bamberg, Stanley W.
Bandalan, Peter C.
Bang, Andrew
Bankert, Roger L.
Banks, Stephen W.
Bankston, Daniel L.
Bankston, James B.
Bannon, George A.
Bannon, William H.
Baran, John T.
Barbee, Walter J.
Barber, George F.
Barber, Jesse L.
Barbudo, Rafael G.
Barcelo, Richard A.
Barcliff, Charles A.
Barcomb, Jamey J.
Bardwell, H. Mark
Barenklau, Keith E.
Bargfrede, Craig A.
Barile, David J.
Barker, James W.
Barksdale, Dane A.
Barnard, Robert G.
Barnes, Andrew M.
Barnes, Everett W.
Barnes, Larry A.
Barnes, Ronnie S.
Barnes, William M.
Barnett, Michael A.
Baronet, Troy D.
Barrera, G. Michael
Barrett, Bruce E.
Barrett, James D.
Barrett, Robert W.
Barringer, John D.
Barringer, Joseph L.
Barron, Michael C.
Barry, Gerald W.
Barry, Michael B.
Bartell, Arthur M.
Bartholdt, William E.
Bartholet, Robert G.
Bartholomeu, Fernando
Bartholomew, Daniel E.
Barton, James L. Jr.
Barttelbort, Larry D.
Bashein, Robert B.
Bashore, James M.
Baskin, Frederick D.
Bass, Harold C.
Bassham, Archie F.
Bateman, Jon F.
Bateman, Sean T.
Bates, Chad T.
Bati, Jose A.
Batiste, Thomas R.
Batschelet, Allen W.
Batson, Bryan K.
Batson, Walter
Battis, Patrick E.
Battle, Jeffrey C.
Battle, Troy Eugene
Battles, Fred C.
Batts, Frank E.
Batts, Jesse J.
Batule, Kevin M.
Bauer, Charlie M.
Bauer, Robert C.
Baugh, William C.
Baughman, Michael L.
Baumanis, Aivars Z.
Baxendale, Gordon M.
Baxter, Leo J.
Bayless, Kevin W.
Bayones, Joseph E.
Baysinger, Howard E. Jr.
Baze, Roy A.
Beach, Gregory E.
Beal, Geordie E.
Beam, Joe C.
Bean, David F.
Beard, Kirby D.
Bearden, John L.
Beardsley, Robert W.
Beatty, John C.
Beaudoin, Guy R.
Becker, Howard A. Jr.
Becker, William A.
Bedrosian, Gregory P.
Bedsaul, Clare D.
Beecher, Robert J.
Beeker, Edward C.
Beers, Roy L.
Beery, Larry N.
Beeson, Randal J.
Beggs, Ivan V.
Beidleman, Robert T. II
Beirne, John E.
Belanger, Van-George R.
Beley, Douglas G.
Belisch, Sigmund B. Jr.
Belkofer, Thomas P.
Bell, Anthony B.
Bell, C. O.
Bell, James C.
Bell, Jesse W. III
Bell, Richard E.
Bell, Robert L. Jr.
Bellisario, Vincent J.
Beltran, Ricky L.
Beltrand, Adrian W.
Bendernagel, Thomas E.
Benevides, Rui C.
Bennett, John W.
Bennick, Gregory A.
Benschoter, Dean L.
Bentley, Allen J.
Bentley, Donald S.
Bentley, Stephen F.
Benton, Chris N.
Benton, H. Frank
Bentz, Jerome R.
Bentz, William J.
Benvenutti, Eugene S.
Beranek, Steve W.
Berberich, Chad S.
Berberick, Raymond E. Jr.
Bergeron, Leo E.
Berggren, David W.
Bergh, Kristopher E.
Bergner, Kevin J.
Bernardi, Roger L.
Bernreuter, Henry G.
Berryman, Craig
Berstler, George A.
Berte, Craig R.
Bertling, Gregory J.
Bertolio, Shane D.
Bessler, Greggary M.
Best, Mark R.
Best, Steven P.
Bethel, Antoine B.
Betker, Robert J.
Beuschlein, Amy B.
Beverley, William W. Jr.
Bevill, Michael D.
Bevington, Richard L. III
Bhalla, Arun K.
Bialozor, William L.
Bialzik, Toby L.
Bieberdorf, Wade J.
Biediger, Paul N. Jr.
Bielanski, Mark J.
Bigelman, Paul A.
Bilafer, Jude P.
Bilby, Richard B.
Biles, Floyd E. II
Billings, Michael B.
Billmyer, John R.
Billot, Mark F.
Billquist, Daniel L.
Bills, Chris P.
Bills, Henry P.
Bills, Thomas C.
Bilo, William C.
Bingham, Leighton P.
Bingham, Raymond L.
Biondi, Richard M.
Birdwell, Brian D.
Bisbee, Dirk D.
Bisno, Chet
Bisping, Thomas A.
Bitting, Alan C.
Bizoukas, Timothy A.
Black, David W.
Black, Donald G.
Black, Frederick H. Jr.
Black, John R.
Black, Nicholas C.

Blackburn, Robert A.
Blacker, Nancy E.
Blackmon, Percy L. Jr.
Blackner, Brad K.
Blair, Christopher T.
Blake, George W.
Bland, William S.
Blejski, James T. Jr.
Blevins, Johnny L.
Bliss, Richard A.
Blodgett, Keith D.
Blohm, Steven A.
Blood, Michael L.
Blood, Nathan B.
Bloom, Eric C.
Blose, Michael D.
Blount, Thomas G.
Blum, Jeffrey A.
Blumberg, Robert C.
Blundon, Sean D.
Blye, Cecil A.
Boatman, Frank A.
Boatner, James G.
Bocchino, Brian A.
Bochenek, Zbigniew
Bode, Glen H.
Boehme, Kenneth L.
Boersma, Frank L.
Bogan, Clyde W. Jr.
Boge, Kenneth G.
Bohnhoff, John A.
Bolen, Thomas R.
Bollinger, Richard H.
Bollinger, Stephen K.
Bolton, Fred C.
Bolton, Patrick C.
Boman, James E.
Bond, Douglas S.
Bond, Gregory F.
Bondra, Craig J.
Bonfoey, Edward M.
Boni, Maurizio
Bonsell, John A.
Bookard, Joe D.
Boon, Thomas J.
Boone, Thomas F.
Boonnag, Chanathip
Booth, John R. D.
Boothe, Stephen W.
Borchers, Michael C.
Borden, Johnny R.
Borges, Carmelo M.
Boring, Jeffrey L.
Boritz, Richard L.
Borreson, Mark W.
Bosher, Boaz
Botts, Stanley E.
Boucher, Glen A.
Boucher, Jeffrey
Bouillion, Kayward J.
Bourgeois, James M.
Bourn, Guy M.
Bourque, Alan G.
Bowen, Crayton M.
Bower, Frederick B. III
Bower, Frederick K. Jr.
Bower, Kenneth A.
Bowers, Michael W.
Bowler, Matthew R.
Bowman, Joseph A.
Bowman, Michael P.
Bowyer, Richard F.
Boyd, Charles R.
Boyd, Morris J.
Boyd, William T.
Boydstun, Carl L.
Boyett, Joseph E. III
Boyington, Richard L.
Boyle, Brian T.
Bozard, John M.
Bozikis, Gerasimos A.
Brackin, John L.
Brackney, Mark D.
Bradbeer, Thomas G.
Bradford, Henry M.
Bradford, Robert D. III
Bradley, Gerald F. Jr.
Bradley, Joseph C.
Bradley, Matthew T.
Bradley, Peter W.
Bradshaw, Larry D.
Bragg, Phillip M.
Braley, Terrence L.
Bralley, Neal H.
Brancato, Anthony G.
Branch, Marion S.
Brandt, Brian J.
Branham, Danny J.
Branstetter, Terry L. Jr.
Brantley, Matthew F.
Branum, Lonnie J. Jr.
Brashear, Bobbie J.
Brashears, David Lee
Braun, Randall S.
Braunbeck, William J.
Bray, Britt E.
Brazell, Joseph L.
Breckenridge, Robert C.
Bredehoft, Brent B.
Brehm, Norman R.
Brehmen, Donald C.
Breitenbach, Daniel L.
Bremer, Mark L.
Brence, Richard E.
Breslow, Edward T.
Bressler, John W.
Brewer, Melvin L.
Bricker, David J.
Brickey, Jonalan
Bridges, Clay F.
Bridges, Joseph W.
Bridgford, Robert S.
Briggs, Robert P.
Brinn, Rufus T.
Briscoe, Mark A.
Brister, Alan A.
Britt, Kenneth W.
Britton, Joseph D.
Broadbent, Robert A.
Broadwater, Colby M. III
Brocious, David E.
Brock, Herbert J.
Brock, Mark E.
Brock, Michael A.
Brodany, William P.
Brodeski, Robert W.
Brodeur, Albert J.
Broms, Donald A.
Broniarczyk, Steven J.
Bronisz, Matthew S.
Bronner, David E.
Bronson, Gary J.
Brooks, Charles W.
Brooks, Joe D. III
Brooks, John C.
Brooks, Jonathan B.
Brooks, Randy L.
Brooks, Roger W.
Brooks, Winston P.
Brostek, Derek J.
Brothers, David J.
Brothers, Lance E.
Brotzen, Paul C.
Brower, Erwin
Brown, Charles E.
Brown, Charles P.
Brown, Charles T.
Brown, David A.
Brown, David B.
Brown, Douglas B.
Brown, Douglas J.
Brown, Gordon C.
Brown, Henry D.
Brown, James A.
Brown, Jeffrey D.
Brown, Jeffrey E.
Brown, Jesse H.
Brown, Joel P.
Brown, John M.
Brown, Larry K.
Brown, Michael E.
Brown, Michael P.
Brown, Peter T.
Brown, Reginald
Brown, Rondall H.
Brown, Ronnie F.
Brown, Ross A. Jr.
Brown, Stanley P.
Brown, Thomas E.
Brown, Thomas J.
Brown, Thomas L.
Brown, Thomas M.
Brown, Titus
Brown, Victor S.
Brown, William
Brown, William C.
Browne, Donald W. Jr.
Bruce, Jonathan A.
Bruce, Steve
Brudzynski, Richard
Bruens, Stephen W.
Bruggeman, Brian T.
Bruha, James J.
Brumfield, Curtis Lee II
Brunner, Richard J.
Bruns, Harold D.
Bryan, Fred C.
Bryant, Alvin Jr.
Bryant, Charles E.
Bryant, Coy K.

Bryant, James H.
Bryant, John H.
Bryant, Marcus D.
Brzozowski, Tom E.
Bucher, Richard T.
Buck, Steven R.
Bucklew, Keith J.
Bugg, Ronnie G.
Bugge, Paul W.
Buhlig, Geoffrey P.
Buhmann, Scott H.
Bullard, Robert L.
Bullard, Tony
Buller, James A.
Bullimore, Steven L.
Bunkley, Grady L.
Bunnell, W. D.
Bunting, Michael S.
Bunting, William C.
Buran, Robert S.
Burdeshaw, William B.
Burgess, LeRoy D.
Burgess, Regnard A. Jr.
Burke, Dennis C.
Burke, Gregory J.
Burke, G. Timothy
Burke, James P.
Burke, John P.
Burke, Patrick F.
Burke, Robert E.
Burmeister, Rodney L.
Burnam, M. Kim
Burnley, Todd W.
Burns, Ferdinand III
Burns, William F.
Burruss, Douglas T.
Bursell, William R.
Burson, Turner I.
Bush, John A.
Bush, Timothy W.
Bussey, Wayne J.
Buster, James A.
Butkera, Edward J. Jr.
Butler, Matthew N.
Butler, Philip R.
Buza, Donald E.
Buzzerio, Leo P.
Bybee, Pearce C.
Bynum, William D.
Byrne, John D.

Cabanatan, Valerian I.
Cabrera, Allan R.
Cady, Cyrus T.
Cahill, William J.
Cain, Dannis L. Jr.
Cain, Robert A.
Cairo, David E.
Caisse, Daniel A.
Calagaz, Frank J.
Callahan, Edwin J. III
Callahan, Robert J.
Callanan, John M.
Callaway, Jack G.
Calloway, Kirk V.
Calvert, Jeffrey A.
Calvin, Matthew L.
Calvo, Angel Luis
Camarella, Dean A.
Camenga, John F.
Cameron, Alan L.
Cameron, Brodnax Jr.
Cameron, Jessie L.
Cameron, John A.
Cammack, Kevin E.
Campany, Eric M.
Campbell, Adam J.
Campbell, Alan E.
Campbell, Albert M. Jr.
Campbell, J. Barton
Campbell, John H.
Campbell, Kelly N.
Campbell, Scott T.
Campbell, Whitney A.
Campbell, William J. III
Canaley, William P.
Cannon, Michael M.
Cano, John A.
Canteel, Brian G.
Canter, John E.
Canterbury, Terry
Cantrell, William K.
Cantwell, Joel R.
Capps, Jack L.
Capps, James S.
Capps, Stephen D.
Capra, Kevin S.
Capra, Robert A.
Cara, Hassen A.
Carafano, James J.
Carboneau, Randall J.

Cardenas, Eduardo
Cardillo, Richard G. Jr.
Carey, Daniel M.
Cariello, Dominic A.
Cariveau, Darrell L.
Carlin, Timothy S.
Carlon, Kenneth G.
Carlson, Timothy S.
Carlson, William J.
Carnahan, Robert K.
Carney, Madison M.
Carns, Timothy A.
Carpenter, Bryan J.
Carpenter, Douglas C.
Carpenter, Gregory S.
Carpenter, Gregory T.
Carpenter, James N.
Carpenter, Randy
Carroll, Kevin J.
Carroll, Marvin
Carroll, Michael J.
Carroll, Steve
Carson, Chester P. Jr.
Carson, Michael T.
Carson, R. Gordon III
Carson, Williams S. II
Carswell, Bruce M. Jr.
Carter, Charles W.
Carter, Duane L.
Carter, Jeffrey C.
Carter, Roy A. III
Cartland, John C. Jr.
Cartwright, Charles A.
Carver, James T.
Casey, Keith A.
Cash, Jack H.
Casillas, Roger A. II
Casner, Craig T.
Caspers, Jeffrey L.
Cassetta, John A.
Cassidy, Thomas M.
Castello, Jason D.
Castells, Paul G.
Castillon, Henry
Castleberry, James D.
Casucci, Jeffrey M.
Catanese, Peter A.
Cate, Hugh C. III
Cates, David L.
Cattell, Don E.

Cearley, Curtis R.
Cecchini, Louis A.
Celestan, Gregory
Cerruti, Lawrence J.
Cervantes, Mario A.
Cesonis, Algis J.
Chabot, Brion V.
Chadwick, Jeffery D.
Chamberlain, Jeffrey P.
Chamberlain, Paul E.
Chamberlain, Wayne J.
Champion, Charles T.
Chance, Kenneth A.
Chandler, Jeffery T.
Chandler, Robert J. Jr.
Chandler, Thomas E.
Chapa, Manuel Q.
Chapman, Gary D.
Chapman, John C.
Chapman, Kenneth E.
Chapman, Thomas M.
Chareat, Paul R.
Charton, Ross W.
Chase, Randall D.
Chase, Vance A.
Chavez, Rocky J.
Cheairs, Terry
Cheatham, Robert G. Jr.
Cheatham, Robert G. Sr.
Cheek, Bryan L.
Cheek, Gary H.
Chelberg, Robert D.
Chenoweth, Charles S.
Cherry, Alfred B.
Cherry, Peter
Cherry, Robert N. Jr.
Cheshire, Walter R.
Chiarella, John G. Sr.
Chiaruttini, Michael J.
Child, Daniel E.
Child, Michael E.
Childers, William A.
Chin, Thomas
Chiste, Ronald L.
Chittivej, Peter K.
Chituras, Jimmie J. Jr.
Chizmar, George A.
Chlebowski, William
Choate, B. F.
Chong, Steve C.

Christensen, Derek
Christensen, Jack R.
Christiansen, Jason K.
Christiansen, Jon F.
Christensen, Richard T.
Christensen, Thomas R.
Christianson, Bernhard E.
Christie, Scott W.
Christopher, Michael G.
Christopher, Warren
Chrzanoski, Stanley V. Jr.
Chun, James H.
Church, James W.
Clampitt, Richard G.
Clancy, Johnson P.
Clapham, David P.
Clark, Brian J.
Clark, Charles N.
Clark, Charles W.
Clark, Christopher W.
Clark, Gregory J.
Clark, James E.
Clark, James R.
Clark, Jeffrey J.
Clark, John
Clark, Joseph P.
Clark, Larry
Clark, Lawrence D.
Clark, Steven L.
Clark, William G. Jr.
Clarkson, Alfred F. Jr.
Clausen, Michael K.
Claysmith, Charles R.
Clayton, Ralph L. III
Cleland, Donald R.
Clem, Robert B.
Clement, Jeanette
Clements, Andrew F.
Clemmons, Philip B.
Cleveland, Gregory J.
Click, Robert L.
Clifton, Jeffrey T.
Clifton, Richard P.
Cline, James D.
Cloninger, Robert L.
Clowous, Freddie A.
Clyde, Tracey
Coakley, Kevin M.
Coan, George P. Jr.
Coats, Wendell J.
Cochran, Marcus A.
Cochran, Roger L.
Cochrane, Charles B.
Cockrell, Truitt L. Jr.
Coe, Robert
Coe, Tony
Coerper, Michael F.
Coffman, Joseph W.
Coffman, Sammy L.
Coghlan, James J. Jr.
Cohen, Elton H.
Cohen, Maurice H.
Colbert, Jason M.
Colburn, Robert H.
Cole, John A.
Cole, John E.
Cole, John L.
Cole, Ronald L.
Cole, Thomas M.
Cole, Timothy P.
Coleman, John E.
Colinsky, Edgar G.
Collier, David K.
Collier, John A. Jr.
Collins, B. F.
Collins, Glen C. Jr.
Collins, Jeffrey C.
Collins, Paul G.
Collins, Richard M.
Colon, Miguel A.
Colon, Samuel E.
Colvard, Eric
Combs, William B.
Compas, Darrell S.
Compter, Thomas A.
Compton, Christopher D.
Compton, Eddie D. III
Condry, Mark E.
Condry, Mathew M.
Cone, Thomas L.
Conery, Paul W.
Congrove, Jack R.
Conley, Charles A.
Conn, Daniel J.
Conn, Roger L.
Connally, Thomas J.
Conner, Tim F.
Conneran, Thomas J.
Connery, Joseph W. Jr.
Connole, Michael S.
Connolly, Kevin G.
Connolly, Michael A.
Connor, John W.
Connor, Paul B.
Connors, Robert E. Jr.
Conrad, Russell G.
Constantine, Nicholas J.
Converse, Douglas M.
Cook, Charles W.
Cook, Christopher W.
Cook, Larry L.
Cook, Ralph J. Jr.
Cook, Samuel C.
Cook, Zale E.
Cooley, Brian M.
Coomler, James D.
Cooney, Norman R.
Coons, Robert
Cooper, Allan E. Jr.
Cooper, Arthur B.
Cooper, Edward V.
Cooper, Paul J.
Copeland, James W.
Copeland, Jeffrey S.
Corby, John F. Jr.
Corda, Mark
Corder, Donald T. Jr.
Cordova, Gonzalo
Corley, Arthur V.
Corley, Michael W.
Cornejo, Robert M.
Cornelio, Frank A.
Cornelison, Lee T.
Cornett, Randall C.
Corpac, Peter S.
Correa, Paul B.
Corville, Michael
Cosgrove, William M.
Cosio, Dino J.
Costella, Ronald G.
Costello, Thomas J.
Costello, William M.
Cote, Joe
Cotham, John A.
Cottle, Mark E.
Cotton, Ross T.
Coull, Jonathan M.
Courson, Michael W.
Courvoisie, Thomas N.
Couture, Roland W.
Couture, Samuel I.
Covington, Johnny W.
Cowart, John D.
Cowell, Craig E.
Cowley, Gerald R.
Cox, Alexander A.
Cox, David M.
Cox, John R.
Cox, Norman J. Jr.
Coy, Eldon L.
Coyle, Thomas M.
Crabtree, Charles F.
Craddieth, Ralph
Craig, George D.
Craighead, Michael R.
Crain, Jeffrey I.
Cramer, Robert G.
Crane, Brian E.
Crawford, Brian A.
Crawford, Danny L.
Crawford, David J.
Crawford, Elton E. II
Crawford, Franklin E.
Crecelius, Gregory J.
Creger, Mary A.
Crespo, Antonio de Vicente
Crews, Dwight G.
Criner, Roger D.
Crittenberger, Willis D. Jr.
Crizer, Scott H.
Crom, Gordon W.
Crone, John A.
Crooks, Rory A.
Crosby, Anthony R.
Crosby, James R.
Crosby, John S.
Cross, David L.
Cross, Michael E.
Crossley, Ross W.
Crosson, Anthony V.
Crow, Stanley E.
Crowder, Jerry L.
Crowder, Ronald G.
Crowe, John T.
Crowell, David V.
Crowley, Patrick J.
Croy, Michael J.
Crozier, William H.
Crum, Ralph M.
Crutchfield, Eddie

Crutchfield, Jerry
Crutchley, Michael J.
Cruz, Arthur E.
Cuff, Michael V.
Cullen, James P.
Cullen, Jason C.
Culver, Joseph L.
Cumming, James M.
Cummings, Donald H.
Cummings, Paul R.
Cummings, Warren S.
Cummins, Ancle W.
Cummins, Eugene L.
Cummins, Gerald E. Jr.
Cundiff, Robert W.
Cunningham, Gregg A.
Cunningham, Jackie L.
Cunningham, James E.
Curfman, Ralph S.
Currie, Matthew W.
Curriera, Carl A.
Curry, Christopher M.
Curry, Kenneth D. III
Curry, Kenneth J.
Curry, Randall C.
Curtin, Michael J.
Curtis, John E. Jr.
Cushing, Jeffrey A.
Cuthbert, Trent R.
Cuthbertson, Scot H.
Cutler, David C.
Cyr, Michael F.
Czaja, Bruce A.
D'Amico, Richard J.
da Silva, Joao Cesar Zambao
Dahl, Kenneth R.
Dahmer, Frank A.
Dalby, Jonathan S.
Dale, Thomas L. Jr.
Daley, Daniel Alan
Dallam, Harold S.
Dalton, Craig L.
Dalton, Jim B. Jr.
Damron, Gerald N.
Dando, Brian J.
Daniel, Charles D. Jr.
Daniel, Clifford R.
Daniel, James P. Jr.
Daniels, John J.
Danikowski, David A.

Danley, John R.
Danner, Richard E. Jr.
Dapore, James M.
Dascanio, Robert A.
Daskevich, Anthony F. II
Datto, John N.
Daugherty, Troy A.
Daugherty, Timothy J.
Daughtry, Clayton M.
Davel, James A.
Davenport, William E. II
Davidson, Thomas E.
Davies, Jesse B.
Davies, Merlyn K.
Davila, Mario
Davis, Bobby G.
Davis, Christopher A.
Davis, Clifford H.
Davis, Cory E.
Davis, Eric M.
Davis, Gregory K.
Davis, James L.
Davis, James R.A.
Davis, Jay S.
Davis, Jeffrey L.
Davis, John R.
Davis, Jon C.
Davis, Jose R.
Davis, Kerry C.
Davis, Michael F.
Davis, Robert J. Jr.
Davis, Robinson C.
Davis, Rodney L.
Davis, Roger W.
Davis, Ronald C.
Davis, Scott F.
Davis, Steven P.
Davis, Thomas Jr.
Davis, Walter F.
Davison, Robert S.
Dawson, Matthew Q.
Dawson, Philip H.
Day, Sean N.
Deal, Paul L.
Dearing, Lamar Michael
Dearolph, Jeffery E.
DeBroux, Daniel A.
DeClementi, Vincent
Decosty, Fiore J. III
Dees, James B.

Defferding, Michael S.
Deichert, Martin K.
de la Plaza, Felipe
DeLand, Clyde G. Jr.
Delano, Douglas C.
DeLarosa, Pedro Jr.
Delerme, Luis F.
Del Favero, David R.
Delgado, Antonio
Delgado, Frankie P.
Delgehausen, Roger D.
Delger, Cory J.
Dellenbaugh, Samuel G.
DeLoia, Matthew A.
DeLong, Allen R.
Del Rosso, Louis J.
Delserone, Ralph P. III
DelVecchio, Michael A.
Demasi, Anthony V.
Demasi, Tony J.
Deming, Duane
Denbow, David L.
DeNezzo, Joseph V.
Denham, Eugene C.
Denker, Roger W.
Dennis, Stephen S.
Dennison, Robert A.
Deranian, Roger C.
Dermon, Robert E.
DeRuyscher, Carl L.
DeShazo, Thomas E.
Desjardins, Lee R.
Detrick, William D.
Dever, Douglas A.
Dever, Eugene A. Jr.
Deverill, Arthur P.
Devine, George K. Jr.
Devine, Patrick L.
Devlin, Stuart A.
DeVries, Lambert D.
DeWalt, Jon A.
DeWeese, Jeffrey L.
DeWesplore, James B.
Dewitt, Gregory P.
Dexter, John N.
Dextraze, Edward G.
Deyeso, Robert L. Jr.
Dezarn, Robert L.
Diaz, Jorge Luis
Dick, Jeremy M.

Dickerson, Gregory B.
Dickerson, Stanley A.
Dickey, James E.
Dickey, Richard A.
Diederich, Andrew R.
Dietz, Andrew L.
Digre, Kenneth B.
DiLillo, Peter T.
Dill, James R.
Dill, John D.
Dillard, John L.
Dillingham, Thomas E.
Dinackus, Thomas D.
Dinges, John A.
Dinoto, Joseph M. Jr.
Dirkse, James W.
Disher, Gregory R.
Ditto, Gordon M.
Dixon, Malcolm R.
Dobson, Douglas B.
Dockery, Dexter R.
Dodge, Andew R.
Dodge, Griffin N.
Dodson, David H.
Dodson, Michael L.
Doescher, Curt W.
Doheny, Michael S.
Doherty, Benjamin N.
Dolgoff, Ross S.
Dolina, John M.
Dollison, Allan L.
Dols, Jonathan R.
Dolton, Harry J. Jr.
Dommer, Joseph T.
Donahoe, Keith A.
Donahue, Michael C.
Donaldson, P. Jeremy
Donnelly, Cornelius J.
Donnelly, John M.
Donohue, Jack E.
Donohue, Kevin S.
Donovan, Michael E.
Doody, Lee A.
Dooley, John P. Jr.
Dorado, Arnel F.
Dorkhom, George H.
Dorow, Alan K.
Dorsey, Matt G.
Dotson, Darryl W.
Dougherty, Kevin M.

Douglass, Robert E.
Dove, Steven W.
Doward, Oscar W.
Downey, Thomas P.
Downing, John L. III
Dowse, Malen E.
Doyle, James L.
Draffen, Terry L.
Drago, John P.
Draper, Douglas G.
Draper, Michael Y.
Drayton, Matthew
Dreibelbis, Scott F.
Driscoll, John J.
Driscoll, Lawrence E.
Drum, Joshua H.
Dubia, John A.
Dublisky, Ernest B.
Duckworth, Alexander E.
Duckworth, Norman E.
Dudek, Daniel J.
Dudney, Lawrence E.
Duffy, Thomas C. Jr.
Dugan, John J.
Dugger, Floyd A.
Duke, Steve
Dumas, Wessley C.
DuMond, John D.
Dumont, Antonio G.
Duncan, John E.
Duncan, Joseph A.
Dunham, John H.
Dunleavy, John F.
Dunn, Charles E.
Dunn, Colin K.
Dunn, Gary J.
Dunn, Jon C.
Dunne, Johnathan P.
Dunnill, William A.
DuRant, Brian D.
Durham, George A.
Durrett, Thomas J. II
Durso, Thomas A.
Dusenberry, Kenneth A.
Dvorshak, Darik D.
Dwyer, John Betpouey
Dyer, Allan G.
Dyer, Hugh N. IV
Dyson, Gregory J.
Dzwonchyk, James D.
Earhart, Douglas B.
Earley, Terry L.
Easom, Grady E. Jr.
Eason, Michael R.
Eason, Robert H. Jr.
Easter, Darrell D.
Easterling, Clay
Eastman, Michael R.
Ebbighausen, John H.
Eckelbarger, Donald E.
Edenzon, Irwin F.
Edge, Steven C.
Edmonds, Peter B.
Edmundson, Thomas J.
Edney, Rachel A.
Edwards, Bryan P.
Edwards, Michael R.A.
Edwards, Robert R.
Egan, Richard T.
Egan, Robert R.
Egger, Robert J.
Ehlers, Bruce A.
Ehlers, Charles L.
Eibert, Edward J.
Eidemiller, Wayne A.
Eimers, Gregory L.
Einerson, David C.
Einwalter, Robert J.
Eisert, Fred J.
Eisiminger, Thomas I. Jr.
Eitner, Edward E.
Elder, Robin L.
Eldridge, David
Elias, Karey L.

Dozers of Company D, 13th Combat Engineers, assisting a self-propelled 8-inch howitzer of 780th Field Artillery up Hill 1181 for direct fire on enemy bunkers, January 1951. (Courtesy of George J. Ellis)

Elkins, James E.
Ell, David J.
Ellerthorpe, Donald C.
Elliott, Gregory C.
Elliott, James C. IV
Elliott, James E.
Elliott, Robert K.
Elliott, Robert L.
Elliott, Tobey S.
Ellis, George J.
Ellis, Hayes G. T.
Ellis, Michael
Ellison, Henry P.
Elsen, David P.
Embrich, Steven R.
Emerson, Charles J.
Emerson, Christopher
Emerson, Marvin E.
Emery, Larry J.
Encarnacion, Conrad J.
Enders, Chad E.
Endres, Michael T.
Enemark, W. A.
Engel, Shawn A.
Engler, Paul D.
English, William R.
Englund, Dale A.
Enneking, Michael D.
Ensslin, Robert F.
Entlich, Richard E.
Erdmann, Donald R.
Erdmann, Howard D.
Erfourth, Montgomery C.
Ergin, Sadettin Alp
Erickson, Eric C.
Erickson, Karl J.
Erickson, Randy J.
Erickson, Vern C.
Erts, Jeffrey G.
Erwin, Charles H.
Erwin, John J.
Erwin, Ray
Escamilla, Christopher R.
Escolopio, Frank A.
Esplin, Mark J.
Esposito, Michael B.
Estes, Franklin J.
Ethridge, Charles B.
Eubanks, Dallas L.
Evans, Alan M.
Evans, David Jr.
Evans, Kris L.
Evans, Michael R.
Evans, Peter H.
Evans, Philip M.
Everly, David R.
Everritt, Wilbur T.
Ewaldsen, Otto A.
Ezuka, Mark K.
Faber, Morris R.
Fager, Arthur B.
Fager, Eric H.
Faherty, Douglas M.
Fahlenkamp, Steven C.
Faint, Arthur J.
Fairbanks, Charles E.
Faircloth, William R.
Fairley, Delbert G.
Falk, Early I.
Falter, Vincent E.
Fandrich, Steven L.
Fant, John S.
Faraci, Charles J.
Farelli, John C.
Farkas, Eric A.
Farmer, Timothy L.
Farnham, Charles
Farquhar, Robert P.
Farr, Philip B.
Farrell, Peter H.
Farris, Joe P.
Farrow, Craig A.
Farwell, Jay E.
Faulk, Curtis W.
Favors, John D.
Fazen, Michael L.
Feagin, Chester C. Jr.
Feaver, Marilyn
Fedak, Peter F.
Feddeler, Douglas J.
Fejerang, John A.
Feldman, Stephen J.
Feliciano, Raul J.
Felix, Kevin M.
Fendig, John P.
Fenn, Mark Herald
Fenoli, Richard M.
Fenton, Gregory P.
Fenton, James P.
Fenwick, Danny M.
Feret, John M. E.
Ferezan, Daniel M.
Ferguson, James
Ferguson, James C.
Ferguson, John R.
Ferguson, Stanley E.
Ferguson, Tim W.
Ferm, Mitchell R.
Fernandez, Stephan J.
Fernandez, Steven A.
Ferrara, Daniel P.
Ferrara, Michael
Ferrell, David D. Jr.
Ferry, Mason
Ferzacca, Albert L.
Field, William E.
Fielding, Fred C.
Fields, Lee J. Jr.
Figueroa, Jesus
Filiatreau, Nezaire A.
Finch, J. Warren
Finck, William H.
Finley, Craig W.
Finley, William J.
Finn, Donald R.
Fiora, Edward L.
Fiore, Anthony J.
Fiorentino, Peter E.
Fischer, Douglas A.
Fischer, Heinrich
Fischer, Hugo J.
Fischer, Nathan S.
Fischer, Robert T.
Fisette, Michael F.
Fisher, David J.
Fisher, James Scott
Fisher, Richard A.
Fisher, Roy L.
Fisk, Robert B.
Fitch, Harlan P.
Fitzgerald, John E. IV
Fitzgerald, Keith E.
Fitzgerald, Michael R.
Fitzke, Burton A.
Flake, Jackson L. III
Flanagan, James A.
Flanagan, Joseph E.
Flatley, Thomas J.
Fleming, A. Jason
Fleming, Joseph A.
Flett, Mark A.
Flieg, David E.
Flint, Charles K. III
Flint, Curtis E.
Flodin, Richard G.
Flores, L. Michael
Flowers, James R.
Flowers, John C.
Flowers, Joseph M.
Floyd, Joe C.
Flusche, Mark S.
Flynn, David S.
Flynn, Francis E.
Flynn, John H. Jr.
Flynn, Mark T.
Fobes, Joanne M.
Foley, Bryan A.
Follas, Thomas J.
Fontaine, John B.
Forbes, Alvis R.
Forbes, Maynard C.
Forbis, Michael D.
Ford, Christopher A.
Ford, Eric J.
Ford, Howard S. II
Ford, Jimmie S.
Ford, Robert B.
Ford, William L.
Forestier, Derrick J.
Forman, Robert H.
FormyDuval, T. L.
Forrest, Michael P.
Forsyth, Michael J.
Foster, Daniel E.
Foster, Duane E.
Foster, John S.
Foster, Michael E. Sr.
Foster, Robert D. Jr.
Foulds, Lloyd A.
Fountain, Reginald L.
Foust, Daniel G.
Fowkes, Harry W.
Fowles, Paul W.
Fowlkes, Jennings B. Jr.
Fox, Glen W.
Fox, John K.
Fox, Sheldon J.
Foy, Robert J.
Frambes, Timothy R.
France, Jimmy E.

Francey, Richard M.
Francis, Alric L.
Francis, Edward J.
Francis, H. Minton
Franell, Joseph E. III
Frank, Kevin M.
Franzetti, Carl J.
Frasca, John D.
Fraser, Robert C. Jr.
Frazer, Ed
Frazier, Jimmie L.
Frazier, Robert A.
Frazier, Robert S.
Freeborn, William J.
Freeman, David M.
Freeman, David W.
Freeman, Randy R.
Freeman, Vincent L. Jr.
Freeman, William L. Jr.
Frego, James P.
Frei, Jason R.
Freund, John R.
Frias, Carlos
Friedly, Douglas E.
Frink, Cecil A.
Frisbie, Gerald M.
Fritz, Allen D.
Fritzler, Rex C.
Fronzaglia, Robert J.
Frost, Geoffrey O.
Frugoli, Francis X. III
Fuellenbach, Shawn M.
Fugate, Dale V.
Fugitt, Bill W.
Fuhr, Leo E.
Fuller, Bradley C.
Fuller, Lonnie B.
Fuller, Ralph E.
Fuller, William S.
Fullerton, Daryl L.
Fulp, Charles L.
Fulton, Christopher T.
Fulwyler, Niles J.
Furber, Daniel L.
Furcolow, David L.
Furney, Richard R.
Furtado, William J.
Futrell, William C.
Gable, Charles L.
Gaddis, Evan R.
Gagnon, Jonathan M.
Gaia, Patrick J. Jr.
Gaillard, John D.
Gaines, Boyd D.
Galarneau, Neil
Gall, William O.
Gallagher, Joe E.
Galland, Mitch
Gallegos, Isaac C.
Gallhaus, Lester T.
Gallick, Joseph J.
Galow, Michael A.
Gambrell, Glenn J.
Gann, Michael J. II
Gansz, Richard J.
Gantt, Kenneth D.
Gapinski, David D.
Garcia, Charles R.
Garcia, Eddie
Garcia, J. Bernard
Garcia, John James
Garcia, Kenneth E.
Garcia, Reinaldo D.
Gardiner, Robert M.
Gardner, Jason G.
Gardner, Jay C.
Gardner, Roger D.
Gardner, Stuart A.
Garfield, Ernest
Garner, John M.
Garner, Kevin L.
Garner, Russell A.
Garren, Terrell N.
Garrett, Benjamin L.
Garrett, Christopher B.
Garrett, Grady B.
Garven, Jack T. Jr.
Garvey, James M.
Gash, Eugene
Gaskell, Douglas M.
Gaston, Joseph E.
Gates, Dannial K.
Gatrell, Gregory S.
Gattis, Floyd D.
Gaul, Robert H. Jr.
Gaumer, David R. Jr.
Gauthier, David T.
Gay, Richard L.
Gaydon, Patrick L.
Gayle, Norwood
Gburzynski, Richard V.
Gearhart, Scott W.
Gearlds, Phillip D.
Geary, Paul J.
Geddy, William B.
Gee, David W.
Gehring, John M.
Geise, Charles E.
Geisenhof, Daniel W.
Geishecker, Alan W.
Geisler, John K.
Geist, Laurence J.
Gelerter, Joshua K.
Gelfand, Michael J.
Gentry, Donald E.
Gentry, Keil R.
Gentry, Mark O.
Gephart, Gregory A.
Geraci, John F.
Gerald, F. Calvin
Gerasimas, Bernard F. Jr.
Gerber, Loyd A.
Gercken, David M.
Gerdes, Kevin S.
Germain, David K.
Germann, Anthony C.
Gerstenlauer, David L.
Getzinger, Philip L.
Ghielmetti, Stefano G.
Gholson, Stephen P.
Giacumo, Robert L.
Gibbs, H. Lee
Gibson, Bruce
Gibson, Stephen C.
Gibson, William H.
Gido, Paul A.
Giebel, Gary D.
Gifford, James M.
Gift, Gary D.
Gilboy, Steven G.
Giles, Leroy
Giliberti, Lucio A.
Gilkey, Michael W.
Gill, David M.
Gill, Jonathan B.
Gill, Matthew T.
Gill, Roger E.
Gillem, Richard D.
Gillette, John W.
Gilliam, Darrel K.
Gilliland, Charles B.
Gillis, Jay N.
Gillotti, Daniel P.
Gilmore, Andrew
Gilmore, Exter G. III
Gilmore, Kenny Bert
Gines, Ernest
Ginter, Rodney K.
Giop, John W.
Giragosian, Aram H.
Girard, Jean
Girard, Marc G.
Gisclon, John L. II
Glacel, Robert A.
Glass, George L. II
Gleason, Robert J.
Gleason, Virgil R.
Glessner, Daniel K.
Glick, John R.
Glisson, Michael J.
Gloriod, John A.
Glowaski, Harvey J.
Glowaski, Michael K.
Goad, Kelly P.
Goecke, Mark A. Jr.
Goettsche, Uwe
Gohsman, Donald C.
Goings, Kenneth W.
Goldfarb, Jeffrey S.
Goldfish, Timothy P.
Goldman, Dwight D.
Goldman, Glenn H.
Goldstein, Harold M.
Golinski, Karl J.
Goltman, John T.
Gomes, Stephen A.
Gomez, Gonzalo A.
Goncz, J. Paul
Gonion, Leo B.
Gonsecki, Jeffery
Gonzales, Julian E. III
Gonzales, Julio
Gonzalez, Ariel
Gonzalez Orriols, Josue R.
Good, Randy S.
Good, Ted M.
Good, Yusef E.
Goode, Mark D.
Goodrich, Peter W.
Goodwin, Frederick C.

Goodwin, Mike D.
Goralnick, Max
Gorden, Fred A.
Gorden, Jerry L.
Gordon, Anthony E.
Gordon, Frank G.
Gordon, Stanley J.
Gorgas, Fred
Gorrell, Timothy C.
Gorski, Frank J. III
Gorsky, Nicholas
Gorzynski, Mark A.
Gossard, Robert S.
Goswick, William L.
Gott, Joel T.R.
Gottschalk, Dempsey D.
Gottschling, James A. Jr.
Gotzmann, Robert G.
Gould, Charles
Gourley, J. Leland
Gouvaia, Ronald D.
Govenettio, Kenneth M.
Graber, Tom
Grabowski, Gerald M.
Grabowski, Richard A. Jr.
Grady, Charles M
Graff, George B.
Graft, Thomas O.
Graham, Andrew F.
Graham, Daniel D.
Graham, Mark A.
Gramling, Paul C. Jr.
Granger, Dewey A.
Granrud, Jerome H.
Grant, Dallas C. III
Grant, Kevin N.
Grauel, David W.
Graves, Gary B.
Gray, Calvin E. II
Gray, Christopher L.
Gray, M. Kevin
Gray, Maston L.
Gray, Thomas A.
Gray, Travis B.
Gray-Lewis, Shirley
Grayson, James E. Jr.
Graziotti, Timothy T.
Green, Donald C.
Green, James M.
Green, James R.
Green, Richard B.
Green, Roger C.
Greene, Michael J.
Greene, Steven A.
Greene, Timothy P.
Greene, William N. Jr.
Greenhouse, Paul S.
Greenman, Frederick F. Jr.
Greenup, H. William
Greer, David E.
Greer, John D.
Greer, William L.
Gregg, Harry A.
Gregor, John P.
Gregory, Charles W.
Gregory, William A.
Greig, Hugh S.
Grenald, Gilberto E.
Grenchus, Charles P.
Gresham, Jack N.
Grey, Mark M.
Grice, Michael D.
Grieme, Wayne G. Jr.
Griese, James C.
Griff, Jack
Grifferd, David Jr.
Griffin, John S.
Griffin, John V.
Griffin, John W.
Griggs, Bryan L.
Griggs, Edson T.
Grimes, Forrest M.
Grimes, Forrest M. Jr.
Grindle, Charles E.
Griner, Rhett B.
Grisoli, Frank L.
Groom, Edgar J.
Gross, Charles E. Jr.
Gross, John D.
Grossman, Richard H.
Grubb, Noel D.
Grubbs, Dallas H. III
Grupper, Ira R.
Gruska, Gerald S.
Guarriello, David M.
Guelle, Paul E.
Guenthner, Jack E.
Guernsey, Jeff R.
Guerra, Jesus
Guerrero, Julian
Guerry, Charles J.
Guertin, Dennis D.
Guffey, Howard R.
Guidos, John M.
Guimond, Troy E.
Gullett, Brett J.
Gullo, Roger A.
Gundlach, Fred W.
Gunhus, Leif W.
Gurr, Rickie C.
Guthrie, John R.
Guthrie, Michael J.
Gutierrez, Carlos
Gutierrez, Robert A.
Gutting, Lucien G.
Guzzo, Samuel
Gwon, Yi S.
Gyarfas, William J.
Haas, Charles J.
Hackbarth, James A.
Hacker, Joseph M.
Hackett, Brian G.
Haddock, Ralph W.
Haffey, Robert M. II
Hagerman, Russell A.
Hague, Mark D.
Hahn, Christian H.
Hahn, Daniel A.
Hahne, Terry
Haight, John E.
Haines, Joseph C. Jr.
Hajek, Jerald D.
Hakes, James V.
Hale, Harold E.
Hale, John C.
Hale, Matthew T.
Hale, Richard T.
Hales, Robert D.
Haley, Jay W.
Haley, Todd T.
Halfacre, Howard E.
Hall, David W.
Hall, Douglas D.
Hall, Jack Q.
Hall, James E.
Hall, James S.
Hall, John A.
Hall, John D.
Hall, Raymond J.
Hall, Richard M.
Hall, Robert R.
Hall, Russell J.
Hall, Steven A.
Hall, Thomas K.
Hallada, Raphael J.
Hallam, Jay Jr.
Halle, Lawrence R.
Hallgren, Hal E.
Hallman, Andrew L.
Halloran, Joseph E.
Halterman, Jeffrey A.
Haltom, Larry W.
Ham, Thomas B.
Hamann, Charles A.
Hamby, Joel E.
Hamilton, Andrew D.
Hamilton, Eddie L.
Hamilton, Jimmie L.
Hamilton, John A. Jr.
Hamilton, Matthew T.
Hamilton, William H.
Hamlin, Wayland
Hamm, Kevin T.
Hammann, William T.
Hammar, George L. IV
Hammes, Tony J.
Hammond, James W. II
Hammond, Thomas M.
Hampton, Dana Neil
Hampton, David R.
Hancock, Jeffery L.
Hancock, Thomas J.
Hanes, Robert E. II
Hanesalo, Bruce A.
Hanesworth, Richard F.
Hanhauser, George J. IV
Hanke, Lorenz J.
Hanks, Harold S.
Hanley, James P.
Hanley, John C.
Hanley, Robert M.
Hanna, Christopher J.
Hanna, Richard M.
Hannah, Bruce James
Hannigan, Edward J.
Hanrahan, Dennis L.
Hansbarger, Thomas D.
Hanscom, Steven M.
Hansen, Henry M.
Hansen, John L.

Hansen, Louis J.
Hansen, Michael B.
Hansinger, Thomas R.
Hanson, Kevin M.
Hanson, Mark D.
Hardimon, James A.
Hardin, Billy V.
Harding, Douglas A.
Hardy, William C.
Harley, Nathaniel Jr.
Harmon, Garrick M.
Harmon, Larry
Harmon, Sean M.
Harmon, Thomas C.
Harmon, William T.
Harnly, Richard W.
Harnois, Lois J.
Haro, Joseph A.
Harper, Alan R.
Harper, Gerald F.
Harper, Neal W.
Harper, Robert L.
Harpster, Deborah A.
Harraghy, Stephen J.
Harrell, Ned L. Jr.
Harrell, Steven V.
Harrelroad, John P.
Harrigan, Nathan S.
Harrington, Joseph P.
Harris, Amani Y. N.
Harris, Anthony J.
Harris, Anthony N.
Harris, Brenda L.
Harris, Bruce S.
Harris, Carsten M.
Harris, Clifton C.
Harris, Elwood O.
Harris, Gerald D.
Harris, John R. II
Harris, Mark P.
Harris, Reginald M.
Harris, Richard L.
Harris, Robert A.
Harris, Rodney
Harris, Thomas E.
Harris, Volmer W.
Harrison, Benjamin A.
Harrison, George H.
Harrison, Wayne C.
Harrison, Woodrow W.
Hart, John D.
Hartley, David T.
Hartman, Dennis R.
Hartmann, Geoffroy
Hartsell, Lynn J.
Harvey, David G.
Harvey, Joe L.
Harvie, David P.
Hassler, John M.
Hastings, C. F. Jr.
Hastings, Michael J.
Hatch, Norman T.
Hatcher, James T. Jr.
Hatton, Phil
Haub, Larry D.
Haubert, John R.
Hauger, Bradley J.
Haugh, David E.
Haught, David D.
Hauschild, John R.
Hauschild, William A.
Hauser, Thomas W.
Hausman, William F.
Havron, Raymond E.
Hawkins, David S.
Hawkins, Walter L. Jr.
Hawko, Thomas J. Jr.
Hawley, Kevin A.
Hawley, Spencer L.
Hawn, Thomas C.
Hawthorne, Dennis R.
Hayashida, David M.
Haycraft, James R.
Hayden, Kevin M.
Hayden, Robert T.
Hayes, Michael T.
Hayes, Rodney G.
Hayth, Bennett E.
Head, Michael W.
Healey, Anthony J.
Healy, James E.
Heard, Michael
Hearne, Willie V. III
Heath, Wayne R.
Hebner, Duane H.
Hecker, William F. Jr.
Hecker, William F. III
Hedeen, Albert J. Jr.
Hedge, Jonathan S.
Hedgespeth, Kenneth W.
Hedin, Glenn C.
Hedman, Daniel K.
Hedrick, Brian K.
Hegerle, Robert J.
Heiberg, William L.
Heidecker, Steven P.
Heidenreich, Donald E.
Heilman, William P.
Heine, Ronald J.
Heinze, Christopher S.
Heitlinger, Gordon
Held, Jason M.
Heldenbrand, Dennis W.
Hellesvig, Ronald E.
Helton, Edward E.
Helvey, Michael A.
Henderson, Austin W.
Henderson, David S.
Henderson, Duane I.
Henderson, Kevin C.
Hendricks, Bert D.
Hendrickson, James A.
Hendrix, Kip A.
Hennes, Mark M.
Hennigan, J. Mark
Hennigan, John R. Jr.
Hennigan, Joseph P.
Henry, Bruce A.
Henry, Henry J.
Henry, Walter E.
Henry, Walter L.
Hensley, Robert B.
Hensley, William R.
Hensley, William R. Jr.
Herbst, Joel
Hercz, Arthur R.
Herd, Karl F.
Herd, W. R.
Hergenroeder, David M.
Hergenroeder, Matthew S.
Herke, Larry J.
Herman, Michael P.
Herman, Theodore E.
Hermanson, Patrick M.
Hernandez, Rhett A.
Hernandez, Rodolfo Jr.
Hernandez, Thomas L.
Herndon, Armstead
Herndon, Jeff L.
Herndon, Thomas R.
Herold, Ernest J.
Herring, David H.
Herring, Gregory Keith
Hertel, Vernon W.
Hertel, William R.
Hesco, Paul
Hester, Leroy V.
Hester, Mark
Hester, Robert V.
Hester, Teddy C.
Heston, Bradley A.
Heubach, Erich H.
Heverin, James M. III
Hewitt, Robert A.
Heyman, William
Heyward, Benny L.
Hibbets, Melvin L.
Hickey, Brenda L.
Hickey, James R.
Hickman, Michael V.
Hickman, Ralph E.
Hicks, John M.
Hicks, Kevin C.
Hideo, Miki
Higbe, Wayne J.
Higgins, Patrick J.
Higgins, Warren W.
Hight, Timothy K.
Hightower, Joseph C.
Hildebrant, Todd C.
Hileman, Jerome C.
Hill, Albert Curtis
Hill, David C.
Hill, Jerry C.
Hill, Randolph M.
Hill, Rick E.
Hill, Robert M.
Hill, Steven D.
Hill, Terrence E.
Hill, Thomas J. IV
Hill, Wayne L.
Hill, William M.
Hiller, Fred
Hillison, Joel R.
Hills, Thomas L.
Hillstrom, Tom
Hilson, James W.
Hinchey, Michael J.
Hinde, John E.
Hinds, Thomas R. Jr.

Hines, John G. Jr.
Hinkle, Eddie J.
Hinsberger, Vincent J.
Hinson, John G.
Hinton, James A.
Hirsch, Bruce A.
Hirsch, Gregory J.
Hirsch, Richard D.
Hirschey, Brian K.
Hitchcock, David L.
Hitchcock, Jon E.
Hite, Hampton E.
Hitt, John C.
Hittinger, William R.
Hittle, Russell W. Jr.
Hixson, Scott L.
Hladky, Mark H.
Hladkyj, Yaropolk R.
Hlopak, Edward J.
Hoar, Joseph V.
Hocker, Anthony L.
Hockstra, Gerrit T.
Hodge, Ben K.
Hodge, Harold L.
Hodges, Brian A.
Hodges, Jeffrey G.
Hodgson, Richard R.
Hodnett, James
Hodo, Linnen E.
Hoehne, John A.
Hoerster, Francis J.
Hoffman, John G.
Hoffman, Joseph B.
Hoffman, Thomas R.
Hoffmann, Louis
Hogan, Michael E.
Hogg, Steven D.
Hoggard, James S.
Hogle, Richard D.
Hohosh, James D.
Holbrook, Deric J.
Holcomb, Christopher J.
Holcombe, Marvin C.
Holden, Carl F. III
Holden, Terry L.
Holder, Alex
Holland, Martin J.
Hollander, Richard L.
Holley, Christopher E.
Holley, James P. II
Hollis, James B. Jr.
Holloman, Bruce E.
Holmes, James W.
Holmes, Russell A. Jr.
Holmquist, Richard K.
Holt, Kenneth E.
Holt, Larry L.
Homan, Lance G.
Honea, Michael C.
Honig, Gunter
Honkus, Joseph M.
Hood, Chris A.
Hookness, Robert S. Jr.
Hooper, Andrew S.
Hooper, Russell L.
Hopkins, Christopher L.
Hoppe, Bruce J.
Hopper, Samuel R.
Horab, Joseph S.
Horikawa, Keith K.
Horn, David R.
Horn, James G. Jr.
Horn, Mark E.
Hornbeck, Robert J.
Horst, Kelso W.
Horst, Kelso W. Jr.
Horton, Robert M.
Horton, Rolla M.
Horwitz, David E.
Hoskins, Gil
Hoskins, Kelly Ryan
Hosler, Owen D.
Hosmer, Richard
Hoss, Daniel
Hossack, Timothy C.
Hotalen, William R. Jr.
Houghton, Daniel G.
House, Bruce L.
House, John M.
Houston, Thomas D.
Howard, Carl E.
Howard, Charles E.
Howard, Richard P.
Howard, Timothy J.
Howell, James S.
Howerton, William B. II
Howland, Robert M.
Hoyt, Edward E.
Hribar, Robert S.
Huah, Leow Hsueh
Huang, Albert Y.
Huang, Yun-Te
Hubbard, Todd H.
Hubbard, Zachary P.
Huber, Gary L.
Hubner, Michael W.
Huckfeldt, Matt
Hudspeath, William A.
Huebel, Cliff
Huerta, Carlos C.
Huerta, Fernando M.
Huff, Keith M.
Huff, Timothy E.
Huff, William D.
Huff, William M.
Huffman, Burnside E.
Huffman, James W. III
Huffman, Rex L.
Huggins, Cleveland P. III
Hughes, Daniel P.
Hughes, Edward L.
Hughes, Elbert W.
Hughes, Jeffrey S.
Hughes, John B.
Hughes, Michael
Hughes, Michael P.
Hughes, Philip E.
Hughes, Richard D.
Hugles, Tyrone
Hulkower, Leslie
Hull, Frank R.
Hultquist, David C.
Humberson, Sidney A.
Humphrey, Adam J.
Hungerford, Frank M.
Hunt, Philip D.
Hunter, Howard W.
Hunter, Kirk S.
Hunter, Michael H.
Hunter, Robert D.
Hurst, Douglas R.
Hurst, James T. Jr.
Huss, Joseph M.
Husvar, Christopher J.
Hutchinson, Andrew F.
Hutchinson, Charles F.
Hutton, Jason C.
Hutton, James E.
Huxford, Richard L.
Huxhold, Paul E.
Hymel, Patrick F.
Hynes, Dwayne D.
Iannello, Anthony A.
Ibberson, Steven N.
Ice, Francis E.
Ifflander, David F.
Ikins, James F.
Illi, Gary M.
Imfeld, Douglas A.
Ingram, Bernd L.
Inhofe, Jim
Inman, Paul M.
Inouye, Robert N.
Irwin, David F.
Isaacson, Avram J.
Isby, David C.
Ivanoff, Kelly W.
Ivester, Terry A.
Jablonski, Edward S.
Jacks, Holger
Jackson, Andrew L. Jr.
Jackson, Charles W.
Jackson, Daniel T.
Jackson, Eleazar J.
Jackson, James G.
Jackson, L. Jay
Jackson, Lonnie C.
Jackson, Paul T.
Jackson, Richard M.
Jackson, Shawn Q.
Jackson, Stephen C.
Jacobi, Thomas M.
Jacobs, Walter S.
Jacobsen, Mark E.
Jacobson, Eric J.
Jacobson, Russell D.
Jagoe, Marcellus H. IV
James, Arnold F. Jr.
James, Brien P.
James, George A.
James, Korya J.
James, Nathaniel
James, Robert D.
Jameson, Charlie R. Jr.
Jancso, Raymond J.
Janey, William H.
Jang, Hyeon Joo
Janiczek, Rudolph M.
Janney, Brooke H.
Janosko, Theodore J.

Jarolimek, Keith R.
Jarosz, Dennis J.
Jarvis, Charles S.
Jaskowiec, Michael A.
Jaszlics, Sheila L.
Javier, Russel P.
Jaworski, Edward A.
Jay, Christopher A.
Jay, Raymond E.
Jeffries, Richard A.
Jeffris, Mark E.
Jelinek, Jonathan G.
Jellison, Rick E.
Jelly, Jeremy E.
Jemison, Paul
Jenkins, Arthur L.
Jenkins, Cynthia G.
Jensen, Arnold W.
Jensen, Craig L.
Jeon, Seung
Jepsen, Robert J.
Jipp, Randall A.
Joel, Kenneth W.
Johansen, Nils I.
Johns, Bradley N.
Johns, Lawrence V.
Johnsen, Nicholas R.
Johnson, Albert Jr.
Johnson, Alex J.
Johnson, Christopher R.
Johnson, David W. Jr.
Johnson, Dean B.
Johnson, Don A.
Johnson, Edward J. Jr.
Johnson, Eric T.
Johnson, Fred Jr.
Johnson, Freeman D.
Johnson, Glenn
Johnson, Gregory D.
Johnson, James
Johnson, James D.
Johnson, James E.
Johnson, James L.
Johnson, James M.
Johnson, Jay D.
Johnson, Jimmie Jr.
Johnson, Kenn W.
Johnson, Mark W.
Johnson, Odell M.
Johnson, Paul L.
Johnson, Randolph R.
Johnson, Raymond E.
Johnson, Robert L.
Johnson, Samuel H.
Johnson, Sheldon D.
Johnson, William D.
Johnson, William M.
Johnson, William R.
Johnson, William T.
Johnston, Kevin L.
Johnston, Warren T.
Jones, Bruce W.
Jones, Bryan R.
Jones, Bryce D.
Jones, Charles
Jones, Craig W.
Jones, Donald W.
Jones, Dwight T.
Jones, James Carroll II
Jones, James E.
Jones, James M.
Jones, Jeffrey W.
Jones, Larry W.
Jones, Lee R.
Jones, Mark E.
Jones, Mark W.
Jones, Michael A.
Jones, Michael M.
Jones, Phillip W.
Jones, Robert R.
Jones, Walter M.
Jones, Warwick N.
Jones, William A.
Jones, William L.
Joo, Kyeong-Sik
Jordan, George H.
Jordan, Glen A.
Jordan, John J.
Jordan, Lee A.
Jordan, Peter D.
Jorling, Joseph H. Jr.
Joyner, Jerold J.
Juarez, Martin R.
Julian, Martin D.
Jung, Javier
Jungerman, George R.
Jurchenko, Dan
Jurden, Delbert A.
Kahler, William C.
Kaiser, Anthony R.
Kaiser, John B.
Kaiser, John J.
Kaiser, William E.
Kalbermatten, Markus
Kale, Donald W.
Kalokerinos, John E.
Kamachi, Keith M.
Kambic, Matthew L.
Kaminski, Mark S.
Kaminski, Michael C. J.
Kampbell, Christopher
Kamper, Kenneth L.
Kandel, Keith
Karinshak, Bryan F.
Karr, Thomas W.
Kascsak, John J. Jr.
Kashaninejad, Roger F.
Kasiski, Stanley A.
Kato, Hisanori
Kauhane, Robert L.
Kaul, Jason
Kawano, Kreiton I.
Keating, Arthur J.
Keating, Peter J.
Keck, Gary L.
Keenan, Robert A.
Keeton, Jerry A.
Keith, Donald R.
Keith, George Allen
Keith, Michael T.
Keller, Jeremy A.
Kelley, Aaron D. Jr.
Kelley, Darryle R.
Kelley, David A.
Kelling, Richard R.
Kelly, Alfred M.
Kelly, Edward V.
Kelly, Evander F. Jr.
Kelly, Gary E.
Kelly, James
Kelly, John E.
Kelly, John J.
Kelly, Joseph P.
Kelly, Kenneth E.
Kelly, Thomas L.
Kelly, William P.
Kemp, Neal W.
Kendra, Joseph A.
Kenison, Coleman D. Jr.
Kenkel, James D.
Kennedy, David N.
Kennedy, James M.
Kennington, Robert V.
Kennison, Richard A.
Keogh, Brian K.
Keppel, Kenton E.
Kerin, James R. Jr.
Kerness, Barry A.
Kerns, Tom
Kerwin, Bruce R.
Kerwin, Walter T. Jr.
Kester, Billy R.
Kesterson, James C.
Ketchum, Kelly R.
Ketron, Neal G. III
Keveles, Dean J.
Keyes, Billy W.
Keyes, Eric E.
Kibe, Josiah L.
Kidd, Gary P.
Kidd, LeRay
Kidner, Martin E.
Kiefe, Joseph C. Jr.
Kiehl, James G.
Kielkopf, Edward C.
Kielpinski, Michael W.
Kilby, Gregory R.
Kilgallon, John M.
Kilgo, Kenneth D.
Killa, Anthony D.
Kim, Dean J.
Kim, Jong H.
Kim, Yong Kil
Kimura, Anson M.
Kinard, Benjamin L.
Kincaid, Richard D.
Kindred, William H.
Kindt, Larry C.
King, Christopher B. Jr.
King, Daniel G.
King, David B. II
King, Ira L.
King, James P.
King, John H.
King, Mark H.
King, Michael J.
King, Peter M.
King, Stanley A.
Kingsbury, Robert P.
Kinman, Bret C.

A Battery, 9th Field Artillery, 155-mm howitzer in action at La Rouge Eaux, France, October 30, 1944. (Courtesy of G. William Mohr)

Kirby, Christopher H.
Kirby, James
Kirby, Robert D.
Kirkland, Robert O.
Kirtley, Frank H.
Kiser, Stephen C.
Kiser, Thomas B. II
Kittrell, James C.
Kiwada, Nobuo
Klaus, Erich
Klein, Jeffrey A.
Kleinfeldt, Daniel S.
Klem, Bruce T.
Klemmer, Keith A.
Kleypas, Kenneth
Klohr, Robert J. Jr.
Kluba, Robert F.
Knapskog, Arne H.
Knauff, James E.
Kneeshaw, Alton S.
Knight, John A.
Knight, Kenneth M.
Knighton, Harry A.
Kniskern, Wayne R.
Knoll, Ulrich
Knott, Arthur W. Jr.
Knowles, Timothy J.
Knowlton, Richard C.
Knox, Philip W.
Koba, Michael G.
Kocsis, Alexander S.
Kodelya, Frank L.
Koehler, Brian D.
Koehler, Collyon S.
Koehler, Gerard J.
Koehler, Robert G.
Koehntopp, Dwayne D.
Koenig, John A.
Koerner, Kevin W.
Koh, Lin-Kee
Kohlhoff, Kris F.
Kokes, Thomas B.
Kolditz, Thomas A.
Kolgat, Joseph Jr.
Kolka, Raymond J.
Koller, Joseph E.
Komar, David P.
Konkus, Timothy G.
Koob, Michael D.
Kool, John E.
Kopko, James P.
Korpal, Eugene S.
Koste, Russell J.
Kotwis, James S.
Kovacs, Peter C.
Koyle, Jeffrey L.
Koziar, William J.
Kraft, Dennis R.
Krahn, Gary W.
Kramer, Jeffrey J.
Kramer, Peter W.
Krasinski, Robert S.
Kratzer, David B.
Kratzer, James W.
Krebs, Christopher M.
Kreider, Stephen D.
Kreye, John R.
Krieger, Mark P. Jr.
Krohn, Marsha J.
Krondak, William J.
Krows, Gerald F.
Kubista, Theodore S.
Kucksdorf, Richard W.
Kuhn, Thomas N.
Kulhanek, Ethel K.
Kulick, Carl M.
Kuntz, Charles C.
Kunst, Michael J.
Kwong, Stan
Kyle, Ronald K. Jr.
LaCagnin, Scott E.
Lacey, Raymond P.
Lackovic, Christopher J.
Lacquement, Richard A.
Lacy, Kyle G. Sr.
Ladman, Bruce F.
Laffosse, Miguel D. Jr.
Lafleur, Peter J.
Lahti, Eric R.
Laing, Richard A.
Lake, Corey L.
Lake, Gordon C.
Lakin, Charles P.
Lally, Paul D.
Lamar, Brian R.
Lamb, Michael T.
Lamb, Steven F.
Lambert, James L.
Lambesis, Christopher
Laminsky, Edward A.

Lamm, Chess P.
Lamonhagine, Paul F.
Lance, Joseph M. III
Lance, Zane A.
Land, John E. Jr.
Landree, Michael L.
Lane, Andrew S.
Lane, Charles D.
Lanford, Richard L.
Lang, Edmund H.
Lange, Gary W.
Langland, Richard F.
Langley, Jeffery E.
Langston, Gregory
Lanier, Larry D.
Lank, Joseph M.
Lapekas, Stephen J.
LaPorte, Ryan J.
Lark, Philip S.
Larkin, Douglas D.
Larkin, Kevin L.
Larkin, Robert J.
LaRock, Harold L. II
Larrabee, Christopher L.
Larrabee, Jeffrey C.
Larsen, Craig M.
Larsen, Henry S. III
Larsen, Robert F. Jr.
Larson, David W.
Larson, Norman H.
Larson, Tod
Larson, Tod M.
Lascher, Frank
Laslie, James C. III
Latendresse, Raymond M.
Lattanze, Christopher A.
Lattimore, Stephen W.
Lattone, Joseph A.
Latz, Robert
Lauck, Stanley M.
Lauer, Zachary C.
Laughlin, J. B.
Laughlin, Kelly D.
Laughlin, Kevin T.
Lauritzen, Kelly S.
LaValley, Stephen M.
Law, Gregory M.
Lawrence, Daniel W.
Lawrence, Jeffrey
Lawrence, John M.
Lawrence, Kevin
Lawrence, Timothy S.
Lawson, David R.
Lawson, Roger
Lazaro, Javier
Leach, Matthew L.
Leach, Robert Blake
Leahy, Brian J.
Lear, Tracy L.
Leard, Robert E.
Lechowich, Richard A.
Ledbetter, Dain C.
Lee, Bryan L.
Lee, David A.
Lee, David S.
Lee, Hoon Ki
Lee, Howard E.
Lee, James W.
Lee, John C.
Lee, John J.
Lee, Kenneth I.
Lee, Patrick T.
Lee, Terry M.
Lee, Thomas J.
Leehneer, Paul R.
Leeman, Sean M.
Lefette, Timothy H.
Legerski, Kevin J.
Lehman, Johnathan H.
Lehman, William P.
LeHockey, John D.
Leigh, Joseph J. Jr.
Leija, Francisco J.
Lemak, Lucian J.
Lemanski, Raymond E.
LeMieux, Guy F.
Lemm, Frederick S.
LeMoine, Christophe
Lennox, William J. Jr.
Lenox, Chester A.
Leon, James
Leone, Louis C.
Leonhardt, Michael E.
Leonick, Gerald J.
LePage, Paul W.
LePage, Phillip W.
LePori, Kenneth B.
Lerner, Jason
Leroux, Gary C.
Leskovac, Thomas J.
Letzring, Steven J.
Levine, Seymour L.
Lewandowski, Timothy J.
Lewey, Terry E. Jr.
Lewis, Alan B.
Lewis, Allen D. Jr.
Lewis, Brett G.
Lewis, Carl Dean
Lewis, David R.
Lewis, Gary C.
Lewis, Harry E.
Lewis, John W.
Lewis, Thaddeus T.
Lewis, Thomas W.
Lewis, Vernon B. Jr.
Lewis, William I. Jr.
Lichtenberger, Robert N. Jr.
Liddell, David J.
Liddy, Robert F.
Lieb, John L.
Liebe, Richard H.
Light, Gregory S.
Lightner, Thomas G.
Liivak, Heldur
Lillard, Robert L.
Lincoln, James B.
Lindell, Scott R.
Linden, Rudy
Lindon, Mark R.
Lineberry, Jonathan G.
Lipscomb, James H. III
Lisak, L. A.
Litterini, Norman P.
Little, Christopher W.
Little, Richard S.
Livingston, Richard M.
Lloyd, Cecil N.
Lloyd, James F. Jr.
Lobash, John A.
Lober, Christian K.
Loch, David H.
Lock, Alan
Lockard, William M.
Locken, Mark A.
Lockett, Carl P.
Loewen, Edward D.
Loftus, Cory L.
Logsdon, Harold K.
Loidolt, Neal G.
London, Douglas T.
Long, Jeffrey W.
Long, Rod
Long, Russell W.
Longarzo, Michael L.
Longo, Richard C.
Lono, Ruddy K.
Loomis, John C.
Looney, James J.
Lootens, Philip C.
Lopez, Chris J.
Lopez, Fernando M.
Lopez, Jose M.
Lopiano, Patrick L.
Lord, Bradford J.
Lorenz, Dwight L.
Lorkowski, Donald L.
Losekamp, Bernard F.
Losey, Wayne E.
Losik, Thomas G.
Loucks, Irvan K.
Love, Christopher J.
Love, Gene
Love, Geoffrey E.
Lovoie, Normand P.
Lowery, Charles E. Jr.
Lowery, James E.
Lowrey, Patrick R.
Loy, Kevin J.
Loy, Sherman A.
Lubold, Guy M. Jr.
Lucas, Angel L.
Lucas, David W.
Lucas, Leroy E. Jr.
Lucente, Dennis R. Jr.
Lucero, Steven P.
Lucke, Daniel L.
Lucynski, John A.
Ludowitz, Robert H.
Ludwikoski, Edward P.
Luebke, Ronald A.
Luedtke, Brian T.
Luhring, Timothy R.
Luitje, Thomas M.
Luker, Mark D.
Luley, Brian J.
Lull, Kenneth J.
Lulos, Nicholas A.
Lund, Dale B.
Lund, James R.
Luper, Benjamin R.

Lupher, John E.
Lupo, John E.
Lushbaugh, Conrad F.
Luther, John J.
Lutkenhouse, John F.
Lutz, Stephen M.
Luzadder, Edward W.
Lyerly, John S.
Lyman, Richard H.
Lynch, Stephen L.
Lynge, Chester J. III
Lynn, Clark III
Lynn, Daniel G.
Lynn, William R.
Lyons, Gerard J.
Lyons, Ralph J.
Lyons, Richard D.
Lyttle, Brian J.
Macak, Richard J. Jr.
MacBride, Roderick J.
MacGillis, James J.
MacGillis, Patrick J.
MacGregor, James W.
MacGregor, John J.
Machen, Robert K.
Maciel, Mark D.
MacIntyre, Bruce C.
Mack, Andrew W.
Mack, Charles C.
Mack, Kenneth L.
MacLean, Mark S.
MacLeod, Sterling D.
MacMillan, Jack F. Jr.
MacNab, Steven
Madden, Michael T.
Madison, Todd L.
Madonna, Victor D.
Madrid, Stephen C. II
Maender, Ryan O.
Magee, Michael P.
Magee, Robert K.
Maggelet, James T.
Magnusson, Michael B.
Mahoney, Michael T.
Mahoney, Paul L.
Mahyudin, Dahlan Jr.
Majancsik, Richard A.
Major, Melvin M.
Malay, John P.
Malherek, Patrick J.
Malicki, Gregg H.
Malloy, John M.
Malmgren, Jim E.
Malone, Dennis M.
Maloney, Michael J.
Malott, David E.
Mammay, Michael J.
Mandel, Marc A.
Mangus, Nicholas M.
Manion, Robert L. Jr.
Manka, Cory S.
Manners, Patrick M.
Mannon, Michael G.
Manring, Gordon M.
Mansell, Roger
Manti, Thomas P.
Manuel, John F.
Manwarren, Howard L.
Maples, Michael D.
Marallos, Christopher Q.
Maranian, Stephen J.
Marble, Andrew D.
Marchi, Randall R.
Marcinkus, A. Frank
Marholz, Duane J.
Marinkovich, Sava C.
Markarian, Ronald H.
Markovetz, Robert F. Jr
Markwell, Brian D.
Marlette, Jeffrey P.
Marlow, Johnny Neal
Marosok, James E.
Marquardt, James
Marrah, Richard W.
Marsh, Brian L.
Marsh, Cody L.
Marsh, Kyle J.
Marshall, George E.
Marshall, Herschel W.
Marshall, James M.
Marshall, Malcolm
Marshall, Trey A.
Martin, Ambro
Martin, Daniel P.
Martin, David A.
Martin, James R.
Martin, John T.
Martin, Larry E. Jr.
Martin, Richard C.
Martin, Stephen E.
Martin, Steven R.
Martin, Van N.
Martinez, Matthew M.
Martinez, Mauricio E.
Martinez, Samuel
Martinez Ferrer, D. Jose Maria
Martini, Kim J.
Marty, Fred F.
Marty, F. Michael
Martz, Randolph P.
Maryott, Randy
Masch, Don
Masiello, Curtis M.
Mason, Raymond E.
Massengale, Stanley C. II
Massey, George S.
Massey, Ronald F. Jr.
Mast, Jack H. Jr.
Mast, Terry L. Jr.
Masterson, James M.
Masterson, John H.
Mastick, Matthew G.
Matakas, David Victor
Mathiesen, Richard C.
Mathieu, Jerome J. Jr.
Matison, Michael J.
Matthew, Thomas E.
Matthews, Marcus
Matties, James P.
Mattingly, James R.
Mattox, James A.
Mauga, Jason R.P.
Mauldin, Nick S.
Maultsby, David A.
Maurer, John A.
Mauss, Irene G.
Maxton, Gregory C.
Maxwell, Eugene P.
Maxwell, Robert E.
May, James D.
May, William R.
Mayette, Christopher T.
Mayeur, Russell A.
Maynard, Richard L.
Mayo, George Jr.
Mayock, Thomas J. Jr.
Mays, Sterling R.
Mazza, Meredith
Mazzuca, Paul Jr.
McAlpine, Donald E.
McAnelly, Gene C.
McArdle, Kevin J.
McAuley, John M.
McAuliffe, Leslie Ann
McAvoy, Craig F.
McBride, Don'll A.
McCadden, Charlie R.
McCahan, George R. III
McCall, Daryl K.
McCandless, James T.
McCarthy, Albert J. Jr.
McCarthy, Brian J.
McCarthy, Erin
McCarthy, John N.
McCarthy, Mitchell J.
McCauley, David J.
McCauley, Kenneth L.
McCay, James R. Jr.
McClain, Michael A.
McClary, James F.
McClelland, John S.
McClung, Charles M.
McClung, Clarke M.
McClure, Kent R.
McCollum, Earl E.
McConachie, Andrew J.
McConnell, Andrew F.
McConnell, John R.
McConnell, Kevin M.
McConnell, Richard A.
McCord, Mark T.
McCorkendale, Christopher
McCormac, Weston A.
McCormack, R. Brian
McCormick, Daniel J.
McCracken, Charles E.
McCracken, John J.
McCrary, Earl C.
McCulley, David E.
McCulloch, Frank W.
McCullough, Brian R.
McDaniel, David S.
McDaniel, Jerry T.
McDaniel, John C.
McDaniel, Lance A.
McDannald, Donald J.
McDermott, William J.
McDonald, Brian C.
McDonald, Eugene P. Jr.
McDonald, James M.

McDonald, John A.
McDonald, Kenneth D.
McDonald, Robert M.
McDonnell, William E.
McDow, Michael T.
McFarley, George Jr.
McFarren, Freddy E.
McGandy, Michael D.
McGee, Michael P.
McGehee, Robert O.
McGill, Jack
McGown, Gary L.
McGraw, Donald C. Jr.
McGuinness, William D.
McGuire, Curtis E.
McGuire, Jeremy P.
McGuire, Mark A.
McGuire, Thomas J.
McGurk, Terrence J.
McHugh, Stephen J.
McIntyre, Brian L.
McKean, Paul G.
McKee, Christopher S.
McKeel, Thomas P.
McKeeman, Michael W.
McKenzie, Eugene A. Jr.
McKeon, Joseph M. Jr.
McKeown, Troy D.
McKiernan, Brendan E.
McKiernan, Brian J.
McKinley, C. H.
McKinley, Matthew R.
McKinney, Paul D.
McKitrick, John H.
McLain, John H.
McLain, Willis A.
McLaughlin, Lester A.
McLawhorn, Linwood E.
McLennan, Stuart G. III
McLeod, Allen O.
McLeod, David
McLeod, William E.
McMahon, Mrs. John J.
McManus, Norbert R.
McMeen, Scott R.
McMillan, Bruce
McMillen, James P.
McMillian, Donald G.
McMullan, James L. Jr.
McMullen, Kieran E.
McNally, Keith
McNally, Patrick K.
McNally, William R. IV
McNamara, Thomas M. Jr.
McNamara, William T.
McNary, Reid W.
McNeilly, Henry I. B.
McNerney, Brian C.
McNerney, John C.
McPhaul, John H. Jr.
McShane, David C.
McSherry, J. L. III
McWhorter, Clint S.
Mead, Robert B.
Mead, Scott E.
Meador, Danny L.
Medcalf, Ruth F.
Medd, Timothy M.
Medof, William M.
Meek, Gary L.
Meek, Martha J.
Meese, Michael J.
Megofna, Anthony J. P.
Mehrer, Bert W.
Meinshausen, Brian A.
Mekelburg, Scott A.
Melancon, Louis P.
Melancon, Steven L.
Melanson, Robert A.
Melcher, Ralph W.
Melin, Andrew
Mellor, Glenn O.
Mellor, John L. Jr.
Mellskog, Donald J.
Melton, Donald R.
Melton, Douglas W.
Melville, Edmond K.
Melvin, Matt G.
Memmola, Lawrence K.
Meno, Brandon G.
Menti, Robert L.
Mentzer, Rodney A.
Menzoff, Julius A.
Mercado, Roberto J.
Mercuri, Joseph W.
Mergele, Jason M.
Merrell, Richard S.
Merrick, Herbert F. Jr.
Merrick, Matt R.
Merrill, David F.
Merrill, Joseph B.
Merritt, Jack N.
Merritt, Michael D.
Merryman, David W.
Mervine, Thomas R.
Messick, Michael K.
Messmore, Harry L. Jr.
Metayer, Joseph A.
Metcalfe, Arthur S.
Mette, Christopher W.
Metz, Philip J.
Metzer, Steve
Meyer, Calvin H.
Meyer, Dennis
Meyers, Berri K.
Miagany, Ricardo
Michael, Eric P.
Michaels, Terry P.
Michalski, David M.
Michaud, John C.
Michel, Christopher J.
Michel, Gabriel J.
Michitsch, John F.
Middleton, Douglas J.
Midyette, Jack B. Jr.
Miesner, Kurt A.
Mikitish, Michael J.
Miklos, Michael T.
Milano, James M.
Milas, Steve
Milelli, Joe
Millar, Hugh L. C. Jr.
Millar, John P.
Millar, Lee G.
Miller, Arthur D.
Miller, Carlos M.
Miller, Charles L.
Miller, Cynthia A.
Miller, Daniel G.
Miller, David B.
Miller, David E.
Miller, Donald Reed
Miller, Frederick L.
Miller, Gary W.
Miller, Geoffrey D.
Miller, Gregory A.
Miller, James L.
Miller, James M.
Miller, James R.
Miller, Jeffrey A.
Miller, John M.
Miller, Johnny R.
Miller, Jon
Miller, Joseph B.
Miller, Kevin M.
Miller, Leo P. Jr.
Miller, Lonnie R.
Miller, Marvin A.
Miller, Michael D.
Miller, Michael W.
Miller, Patrick V.
Miller, Richard A.
Miller, Richard G.
Miller, Robert G.
Mills, Carl V. Jr.
Mills, Charles D.
Mills, Christopher D.
Mills, John H.
Milner, Jon R.
Milner, Robert W.
Miltenberger, Joel E.
Milton, Kevin W.
Minervini, Xavier
Minervini, Xavier F.
Minetti, Gerald P.
Minish, Robert B.
Minnich, James M.
Minter, Dwight F.
Minto, John C. II
Minyard, Charles M.
Miracle, William B.
Miraco, Roland N.
Miranda, Luis G.
Mitchell, Edward D.
Mitchell, James P.
Mitchell, Sammie L.
Mitchell, Sean T.
Mitchell, Steven E.
Mitchell, Terry E.
Mitchell, Todd D.
Mitchum, Kevin L.
Mize, Rodnie L.
Moan, Joseph P.
Mockenhaupt, Donald A.
Moeller, Robert C.
Moen, Frank
Moffet, Marion O. Jr.
Mogan, William W.
Moldovan, Horatiu
Molstad, Ronald O.

Moltz, Michael W.
Monaco, Francis J.
Monday, Joseph A.
Monk, William III
Monkman, Bruce N.
Monroe, Grover C.
Monrreal, Daniel
Montague, Kareem P.
Monteverde, Miguel E. Sr.
Montgomery, Douglas J.
Montgomery, Erby H. Jr.
Montgomery, Glenn A.
Montoya, James H.
Moody, Kevin L.
Moon, Brad A.
Moore, Alan L. Jr.
Moore, Charles R.
Moore, Darren C.
Moore, David R.
Moore, Edmund W. III
Moore, Jeffrey L.
Moore, Jeremy B.
Moore, Kenneth M.
Moore, Lance A.
Moore, Michael A.
Moore, Michael C.
Moore, Richard H. Jr.
Moore, Ronald Patrick
Moore, Russell A.
Moore, Stephen E.
Moore, William R.
Moosmann, Christopher P.
Morales, Julio V.
Morales, Luis G.
Moretti, Brian E.
Moretti, Christopher S.
Moretti, Luke J.
Morey, William P.
Morgan, Gary D.
Morgan, John W. III
Morgan, Kenneth
Morgan, Kerry J.
Morgan, Nathan B.
Morgan, Thomas D.
Morgigno, Raymond C.
Morig, Robert D.
Morin, Carl R.
Morris, Albert C.
Morris, Brett E.
Morris, Bruce
Morris, Carroll N.
Morris, Charles W.
Morris, David D.
Morris, Michael F.
Morris, Scott B.
Morris, Todd B.
Morrison, Mark L.
Morrison, Paul W.
Morrison, Roger A.
Morriss, Philip J.
Morton, David A.
Morzenti, David C.
Moseley, Charles Lee
Moseley, James W.
Moseley, Michael Bruce
Moser, John W.
Moses, George L.
Moss, Alan M.
Moss, Francis R.
Moss, Johnnie D.
Moss, Stanley B.
Moughon, James C. III
Moulton, Royal J.
Mount, Steven R.
Mowchan, John A.
Moxley, Harold L.
Moyer, William R.
Mrozek, Albert A. Jr.
Mueller, Mark R.
Mueller, Warren D.
Mues, Lloyd E.
Muhlenfeld, William F.
Muldrew, Anthony
Mullek, Paul J.
Muller, August
Mullett, John A.
Mullins, Michael S.
Mumpower, William M.
Munden, Ron
Mundt, Kenneth E.
Mundy, Steven E.
Munsey, Thomas E.
Munster, Conrad H. Jr.
Murati, George J.
Murdock, Edward T.
Murdock, Gordon G.
Murphy, Dennis M.
Murphy, Francis P.
Murphy, James J.
Murphy, Kevin P.
Murphy, Mark S.
Murphy, Richard L.
Murphy, Thomas J.
Murray, Andrew J.
Murray, John D. II
Murray, Martin S.
Muse, John D.
Muzzy, Richard H.
Myers, Frank T. II
Myers, J. Blake
Myers, Leslie Clifford
Myers, Neil D.
Myers, Stephen C.
Myrick, Paul R.
Myslinski, Thomas M.
Naas, Robert J.
Nach, David L.
Nagler, Michael R.
Nakamura, Masashi
Nakopoulos, Georgios P.
Nall, Jeffrey M.
Nall, William C.
Nannini, Vance J.
Naples, Anthony J.
Napoli, Joseph F.
Nardone, Ralph
Narimatsu, Jon M.
Nash, Charles W.
Nazario, Edgar D.
Neal, John J.
Neason, Clarence Jr.
Neely, William V.
Neff, Casey A.
Neiggemann, William D.
Neilitz, Mike J.
Neitzel, Robert J.
Nelson, Dale G.
Nelson, Donald S.
Nelson, George H.
Nelson, Lincoln S.
Nelson, Neil E.
Ness, Martin K.
Nester, David C.
Nesvik, Brian R.
Nettles, John G.
Nettleton, Steven W.
Neuendorf, Thomas D.
Nevens, Langdon G.
Nevin, Harold J. Jr.
New, Richard C.
Newbry, Everett G.
Newcomb, Douglas B.
Newcomb, Sammie D.
Newell, Todd F.
Newkirk, O'Neil J. Jr.
Newman, Randall H.
Newton, Francis C. Jr.
Newton, James S.
Neylon, James B.
Ng, Gary C.
Ng, Richard
Nicholas, Robert M.
Nichols, Harold K.
Nichols, James W.
Nickerson,Edward A.
Nickles, Don
Nicks, Eric O.
Niederriter, Craig A.
Nielsen, Kirk M.
Nieminski, John J.
Nieves, Luis F.
Nipp, Robert F.
Nizolak, Joseph P.
Noble, Earl D.
Noble, Scott E.
Nobles, Charles S.
Noel, John M.
Nolan, Daniel A. III
Nolan, Scott M.
Nolan, Scott P.
Noll, Steven R.
Norberg, Joseph M.
Norberg, Seth A.
Nord, Arne E.
Nordrum, John C. Jr.
Norman, Duane M.
Norman, Jere P.
Norman, Joel W.
North, Taw D.
Norton, James F. Jr.
Norton, Michael J.
Norton, Timothy P.
Norway, Robert M.
Nosek, Edward
Nossov, P.
Novak, Brent E.

Novotka, Walter M.
Nowacky, Elliott C.
Nowak, Keith M.
Nowak, Thomas V.
Noyes, James L.
Noyes, John S.
O'Callaghan, Joseph M.
O'Connell, Peter
O'Connor, Chris E. Jr.
O'Connor, Christopher J.
O'Connor, Michael E.
O'Connor, Olen R.
O'Dell, Steven O.
O'Donnell, Thomas J.
O'Hara, John M.
O'Hara, Patrick
O'Hare, Russell J.
O'Hearn, Kevin C.
O'Keeff, Dennise J.
O'Lear, Frank P.
O'Mara, David W.
O'Neal, Jeffrey T.
O'Neil, David P.
O'Reilly, Timothy L.
O'Rourke, Terence J.
O'Sullivan, Curtis Hooper
Oatman, Steven L.
Ocha, Montie J.
Odell, John C.
Odierno, Raymond T.
Oedekoven, Olin O.
Ogle, Robert E.
Oglesby, Paul L. Jr.
Ohl, Walter R.
Olejasz, Stephen N.
Oliphant, John W.
Olive, Jerome P.
Oliver, Arthur W.
Oliver, Ernest M.
Oliver, James R.
Oliver, Kevin D.
Olivero, John P.
Olmstead, Stephen G.
Olson, Bruce A.
Olson, Gregory A.
Olson, Rodney L.
Olson, Russell V. Jr.
Olson, Steven J.
Olson, Walter E.
Onye, Godfrey E.
Orchard, Michael P.
Orgeron, Herman J.
Orlando, Michael J.
Ormsby, Terence McCabe
Orozco, Juan Carlos
Orrick, Jimmy W.
Ortiz, Victor M. Jr.
Ortiz-Lopez, Gervasio
OrtizVega, Carlos R.
Orzol, John
Osborne, Andrew H.
Osborne, James D. Jr.
Osburn, Norb
Ossorio, Peter M.
Ott, David E.
Ott, David E. Jr.
Otterstedt, Charles C. III
Overby, Lee T.
Overschmidt, Francis S.
Owen, Dale E.
Owen, Dennis R.
Owens, Owen R.
Owens, Timothy V.
Oyler, Francis G. Jr.
Pace, James A.
Page, James M.
Page, John D.
Page, Rod
Page, Sherman D.
Palermo, Frank J. Jr.
Palmatier, Bruce Thomas
Palmer, Charles D.
Palmer, Larry W. Jr.
Palmer, Todd A.
Palmore, Ralph L.
Palumbo, Robert J.
Pampena, Giuseppe
Pangborn, Douglas L.
Papastathopoulos, Stavros G.
Paquet, Raymond E. Jr.
Paquin, Albert J.
Paquin, Donald I.
Parente, John Jr.
Park, Chung Hee
Park, Randall E.
Park, Steve J.
Parker, Kevin Kit
Parker, Kevin S.
Parker, Melvin J.
Parkman, Robert W.
Parks, David R.
Parks, Kenneth W.
Parks, Mark E.
Parnell, William C. III
Parquette, William A.
Parris, Dewey L.
Parsons, Berkeley A.
Parsons, Darren N.
Passet, John R.
Pastore, Dennis N.
Pate, Freddie
Patenaude, Anne M.
Patenge, Robert W.
Patterson, Dan B.
Patterson, Ian
Patterson, James S.
Patterson, John V.
Patteson, Curtis W.
Patton, Christopher A.
Patton, Jackie W.
Patton, Michael S.
Patton, Todd M.
Patty, Allen E.
Patzer, Gary R.
Paul, Gregory J.
Paulo, Walter
Paulson, Bernard
Pautler, Michael K.
Pavlock, Jeffrey J.
Payne, David W. Jr.
Payne, John G. Jr.
Payne, Renwick L.
Paz, Heztor E. III
Peacock, Charles H. Jr.
Pearce, Michael J.
Pearson, Andrew L.
Pearson, Paul F.
Peay, J. H. Binford III
Pecjak, Frank W. Jr.
Peckham, Edward D.
Pedigo, Jeffrey
Pedigo, William E.
Peek, David E.
Peeler, John E.
Pelley, Bryan A.
Peloquin, Edouard A. Jr.
Peltola, Jarmo N.
Pena II, John S.
Penberthy, John M.
Pendell, Jon S.
Pendergast, Mark M.
Pendleton, Thomas
Penney, Loren D.
Penton, Thomas M.
Peoples, Jay C.
Peppmuller, Gerry A.
Pereira, John A.
Perez, Felix M. Jr.
Perkins, Alvin A.
Perkins, Anthony W.
Perkins, Donald R.
Perkins, Elbert N.
Perkins, Jon H.
Perkins, Roy E.
Perrine, John D.
Perrine, John M.
Perrotta, Gregory S.
Perry, Jeffrey C.
Perry, Jeffrey J.
Perry, Michael Roy
Perry, Miller O.
Perry, Robert L.
Perry, Thomas A.
Perry, William J. III
Persson, Kenneth V. Sr.
Peters, Benjamin G.
Peters, David B.
Peters, Frederick Y.
Peters, Roy A.
Peterson, Bryan G.
Peterson, Dean A.
Peterson, Felix Jr.
Peterson, Harry L.
Peterson, James E.
Peterson, John P.
Peterson, Kenneth M.
Peterson, Kevin L.
Peterson, Kevin W.
Peterson, Lance J.
Peterson, Leigh A.
Peterson, Leo Scott
Peterson, Michael C.
Peterson, Paul S.
Peterson, William R.
Petrenko, Victor
Petrich, Gary F.

Petrich, Robert J.
Pettit, Clay D.
Petty, John R.
Petty, Ray
Peveto, Joe C.
Pezold, Harry M. F.
Pfeil, William A.
Pfenning, Mark Alan
Phelps, Parris A.
Phibbs, Jonathan D.
Philbin, Edward J.
Philbin, Michael J.
Philipps, Gavin K.
Phillabaum, Paul B.
Phillips, Charles E.
Phillips, Ilean
Phillips, John W.
Phillips, Kenneth A.
Phillips, Mark A.
Phillips, Michael H.
Phipps, Gregory
Piazzoni, Michael S.
Pichardo, Jermaine A.
Picht, Robert G. Jr.
Pickar, Charles K.
Pickens, Ronald M.
Pickett, Roy L.
Picon, Alfonso J.
Pieper, Kenneth L.
Pierce, Billy L.
Pierce, Brian J.
Piercy, Daniel T.
Pierson, Jeffrey L.
Pigott, John E.
Piirainen, Troy S.
Pike, Eddie W.
Pilgrim, Seth T.
Pino, Eugenio G.
Pitarat, Sakon
Pittman, Tillman
Pitts, Roy E.
Pitts, Russell A.
Placke, Derek W.
Plantz, Rober N.
Plater, Richard C.
Platt, Gregory H.
Platt, Stephen J.
Platten, Peter
Player, Charles M. Jr.
Plitt, Walter E.
Plotner, William F.
Plummer, Jimmy M.
Plunkett, Tony
Pohl, Jack
Pokorny, Anthony G.
Pola, William B.
Polk, Gilbert A.
Pollard, David R.
Pollock, Merlyn J.
Pollock, Mitchell J.
Pollock, Robert M.
Pomije, Henry W.
Pompelia, Dominic E. Jr.
Pon, Jigg T.
Ponce, Benjamin
Pond, Joshua B.
Poniatowski, Edward M.
Poon Ghee, Philip Su
Popa, Thomas A.
Pope, Charles W. Jr.
Porras, Alfred
Porter, Gregory C.
Porter, Melvin E.
Porter, Ray E. III
Porter, Thomas L.
Portigue, Robert J. Jr.
Portman, Terrance M.
Poruban, Kevin M.
Post, Paul T.
Potter, James R.
Potter, Joseph C.
Potter, William M.
Povlich, Lester T.
Powell, Alfred E.
Powell, Beverley E.
Powell, James S.
Powell, John W.
Powell, Rex L.
Powell, Robert L.
Powell, Shane P.
Powell, Thomas C.
Powers, Christopher J.
Prater, Timothy L.
Pratt, John F.
Pratt, Sheldon J.
Preczewski, Stan C.
Predmore, Jeffrey C.
Prenger, Ronald J.
Pressel, Arthur F.
Prestegaard, Dean M.
Preston, Andrew D.
Pribbenow, Merle J.
Price, John E.
Price, Michael L.
Price, Troy M.
Price, Vincent L.
Priegnitz, Warren J.
Prince, George W.
Pritchett, David K.
Pritchett, Thomas Les
Pritzker, James N.
Propes, David N.
Pruiett, Brian L.
Pryor, Philip A.
Pryor, Robert
Ptacek, Steven C.
Puckett, Anthony J.
Puckett, Charles K.
Puckett, Donald C.
Puente, Joseph L.
Puett, Ivan D. Sr.
Purdy, Karl L.
Purrington, Donald E.
Pursell, David L.
Putnam, George W. Jr.
Putz, Michael J.
Pyron, Walter L.
Pyrtle, Michael D.
Quagliata, Richard S.
Quek, Thomas M.
Quigley, Keith B.
Quillen, Madeleine
Quinlan, James A.
Quinn, Carl T.
Quinn, Matthew T.
Quinn, Patrick D. III
Quinn, Richard L.
Quinn, Stephen M.
Quinnett, Robert L. Jr.
Quintero, Thomas W.
Quirici, Russell E.
Raad, Salim M.
Race, Howard C.
Radcliffe, Bernard A.
Radcliffe, Derek O.
Radishek, Walter J.
Rae, James C.
Rafferty, John L.
Ragsdale, James E.
Rahe, Thomas H.
Rainey, Scott A.
Rains, Jeffrey A.
Raithel, Charles F.
Rakers, George J.
Ralston, David C.
Ramirez, Armando
Ramirez, David R.
Ramos, Alfredo
Ramos, Ernesto A.
Ramos, Felix
Ramos, Ramon
Ramsey, Brently L.
Ramsey, Edwin P. Jr.
Ranck, Roger L.
Randolph, Harvey H.
Rannow, Eric C.
Ranson, William G.
Raphael, John D.
Rasmussen, Jay M.
Ratcliff, Craig A.
Ratcliff, Hardy V.
Rauch, Frank C.
Rausch, Timothy J.
Rawlinson, Gordon L.
Rawls, Buddy G.
Rawls, Charles R.
Ray, Michael N.
Raymond, William M. Jr.
Raynor, David B.
Read, Richard D.
Reade, Jeff B.
Reagle, William G.
Reavely, Jack D.
Reddout, Donna J.
Redlinger, Mark J.
Reece, Ralph G.
Reed, Charles W.
Reed, George Jr.
Reed, Hampton
Reed, James
Reed, Jean D.
Reed, Robert E.
Reeder, Ernest
Reeder, Joe R.
Reeder, Redell
Reese, Chester T.
Reese, Robert J.
Reese, Shawn P.
Reeves, David E.
Reeves, William R.

M109A6 self-propelled 155-mm howitzer (Courtesy of the Department of the Army)

Regalado, Rick
Rehermann, Steven T.
Reich, Theodore H.
Reid, Derrick A.
Reifenrath, Francis E.
Reilly, Timothy D.
Rein, Christopher M.
Reiner, Kenneth L.
Reinfurt, Edward J.
Reinhard, Ransford A. II
Reinhardt, Brian C.
Reinhart, William H.
Reis, James M.
Reisweber, Glenn D.
Reither, Wayne C.
Remers, Julian J.
Rendon, Andrew T.
Renfroe, Jerry L.
Rentschler, George A.
Reoyo, Paul J.
Resler, Richard A.
Reukauf, William A.
Reuss, Philip G.
Rexford, Owen P.
Reyes, Rosendo C.
Reynard, Richard L.
Reynolds, John C.
Reynolds, Wallace W.
Rezentes, Sean A.
Rhinehart, Richard A. Jr.
Rhodes, Joseph B.
Riccio, Mark A.
Rice, David K.
Rice, Howard P.
Rice, Jay N.
Rice, Patrick L.
Rice, William D.
Rich, W. Skip
Richard, Maurice
Richards, George M.
Richardson, Benny R.
Richardson, C. J.
Richardson, David E.
Richardson, Duane H.
Richardson, Gordon A.
Richardson, James D.
Richardson, Richard S.
Richburg, Lester E.
Richetts, Daniel A.
Richie, James F.
Richie, Joseph M.
Ricker, Jerome M.
Ridge, John E.
Ridge, Ross E.
Ridgeway, David L.
Ridgeway, Russel
Rigby, Randall L.
Riggio, Frank X.
Riley, Andrew G.
Riley, Christopher
Riley, Jack C.
Ringler, Richard W.
Rinke, Matthew C.
Rinkleff, Stuart C.
Riojas, Jose D.
Ripper, Richard E.
Risberg, Robert H.
Risher, Rhett A.
Rist, William B.
Ristuccia, Angelo
Ritchie, Bruce W.
Ritchie, John A.
Ritchie, Matthew A.
Riter, Andrew C.
Rittenberg, Henry W.
Ritter, Timothy R.
Rivera, Agustin
Rivera, Augie
Rivera, Edwin F.
Rivera, Luis A.
Rivera, Modesto C.
Robards, James A.
Robbins, Christopher K.
Robbins, Glenn B. III
Robbins, James E. Jr.
Robbins, Jason W.
Robert, Arthur P.
Roberts, Daniel S.
Roberts, David J.
Roberts, Jonathan B.
Roberts, Tracey C.
Robertson, Phyllis D.

Robinette, Max G.
Robinson, Alex N.
Robinson, Craig M.
Robinson, Eddie L.
Robinson, Franklin L.
Robinson, Gary T.
Robinson, George M.
Robinson, Jack D.
Robinson, John A.
Robinson, Lindsey R.
Robinson, Mark D.
Robinson, Michael A.
Robinson, Michael D.
Robinson, Thomas J. Jr.
Robinson, Wendell A.
Robison, Bert T.
Robison, Dean E.
Robles, Jose M.
Rochell, Robert N.
Rockwell, Stephen B.
Rodeman, Jeffrey C.
Roder, Mark N.
Rodgers, Charles A.
Rodgers, David W.
Rodier, Edward A.
Rodis, Steve Costias
Rodriguez, Carlos
Rodriguez, Hector R.
Rodriguez, Oscar H.
Rodriguez, Paul
Rodriguez, Robert D. Jr.
Rodriguez, Salvador Jr.
Rodvelt, Gary B.
Roe, Thomas J.
Roecker, Frederick C. III
Roederer, Rodney L.
Rogers, Carter L.
Rogers, David C. III
Rogers, Everett B. III
Rogers, Guy A.
Rogers, Homer C.
Rogers, Howard W.
Rogers, James D.
Rogerson, Charles V.
Rohling, Jeromy L.
Roller, Charles E.
Roller, Lon R.
Romanjenko, Peter A.
Rood, Harold W.
Rookard, Stephen E. Jr.
Rooks, Gregory K.
Rooney, James J. Jr.
Rootness, John M.
Roper, Dan
Roper, Daniel S.
Rorje, Alvie L.
Rosa, Thomas C.
Rosales, Louis E.
Rosdahl, Keith A.
Rose, Frank
Rosen, Melvin H.
Ross, Brian M.
Ross, Douglas M.
Ross, James P.
Ross, Matthew W.
Rossow, Shane L.
Rosten, Gregory A.
Roth, Eric G.
Roth, Robert W.
Roth, Thomas J. II
Rothe, Timothy J.
Rothrock, Kenneth L.
Rottstedt, Vernon R.
Rourke, Robert U. Jr.
Rouse, William I.
Rowan, Thomas D.
Rowe, Michael E.
Rowe, Paul R.
Rowe, Sean T.
Rowe, Terry L.
Rowe, William J.
Rowland, Kevin E.
Rowzee, Fred R. Jr.
Royce, Charles P.
Royer, Robert
Rozenberg, Semyon A.
Rubin, Howard H.
Rucker, Dawn
Rudd, John W.
Rudisill, Guy V.
Rudman, John F.
Rueger, John W.
Ruff, Ronald W.
Ruffin, Rodney E.
Ruhlen, George
Ruiz, Fidel V.
Ruiz, Ricardo
Runey, Michael D.
Runion, Daren
Rush, Chris C.
Rush, Edward J.
Rush, Jim E.
Rush, Robert J.
Russ, Stanley
Russell, Brian E.
Russell, Kim T.
Russell, Terry S.
Russell, Theodore S.
Russell, William M.
Rusting, John A.
Ryals, Kenneth R.
Ryan, John J.
Ryan, Neil P.
Ryan, Robert
Rydell, Terry L.
Ryneska, John J.
Ryvicki, John
Sabatine, Jeffrey M.
Sabia, Giacomo R.
Sachrison, Jon E.
Sachtleben, James L.
Sadiarin, Nestor
Sailors, James R.
Saint Hill, Frederick Jr.
Saint-Pol, Pierre P.
Salazar, Vincent L.
Salie, Brian J.
Salim, Mohd Aris
Sallee, Frank
Sammet, George Jr.
Samulewicz, Marc T.
Sanborn, Jeffrey M.
Sanchez, Manuel
Sanchez-Lailla, Felix
Sandage, James R.
Sanders, Glenn W.
Sanders, Horace L.
Sanderson, Kenneth J.
Sanderson, W. K.
Sandifer, Kelly J.
Sandla, Robert S.
Sandlock, Stanley J.
Sandstrom, David F.
Sanford, Albert G. Sr.
Sanford, Leonard P.
Sanford, Terrance P.
Santana, Pablo
Santiago, Henry
Santoni, Luis G.
Santos, Marcos A Amaro dos
Sarabia, George H.
Sarafian, Gregory R.
Saramago, Julian C.
Sargent, Philip A.
Sartor, Robert L.
Sass, Eric M.
Satchwell, Edgar L.
Satterwhite, Harvey T. Jr.
Saul, Arthur D. III
Saul, Lawrence H.
Saumer, Daniel M.
Sausser, Robert G.
Savell, Chad L.
Savickas, Eric J.
Sawyer, Linwood M.
Sawyer, Michael J.
Sawyer, Samuel G.
Saxon, Robert D.
Sayer, Steven M.
Scammell, Peter J.
Scandrett, Todd M.
Scanlan, John J.
Scarbrough, L. Mike
Schaad, Carl W.
Schafer, Matthew C.
Schafer, Russell G.
Schapel, James P.
Scharp, Robert C.
Schaumann, Kirk W.
Schaus, Rich
Schellingerhout, Thomas
Schemine, Steven M.
Schenck, Douglas A.
Scherer, Clay S.
Schermer, Scott C.
Scheuch-Evans, Lewis
Schevalier, Anthony
Schild, Markus R.
Schiller, James A.
Schindler, Francis H.
Schindler, Gerald R.
Schlatter, Philip J.
Schletti, Ulrich
Schlicht, Robert F.
Schlimgen, Daniel L.
Schloz, Stan

Schmeling, William R.
Schmick, Daniel J.
Schmidt, Christian U.
Schmidt, Eugene W.
Schmidt, James E.
Schmidt, Lawrence A.
Schmitt, Mark E.
Schmitz, Gerald J.
Schmutzler, Gary L.
Schneider, Joseph A.
Schneider, Michael W.
Schofield, Norman M.
Schoomaker, Fred B.
Schosek, Kurt A.
Schottler, Jeffrey L.
Schotzko, John E.
Schrader, Edward C.
Schreyach, Jon C.
Schroeder, Elton
Schroeder, Ron E.
Schroeder, Timothy M.
Schueler, Raymond G.
Schuetz, Gerald D.
Schulte, Daniel J.
Schultz, Gregory B.
Schultz, Richard M.
Schulz, Richard H.
Schutt, Richard T.
Schwamberger, Carl W.
Schwartz, Garry S.
Schwegler, Eric E.
Schwenner, John T.
Scianni, Michael J.
Scott, Charles M. Jr.
Scott, David G.
Scott, James L. II
Scott, Jeffrey J.
Scott, Mark W.
Scott, Steven C.
Scott, Steven T.
Scott, Willard W. Jr.
Scovill, Thomas J.
Scraba, Mark D.
Seaholm, Peter E.
Seamands, Robert
Seamon, John E.
Searer, Charles H.
Seaton, James B.
Seay, Brian A.
Seay, Stephen M.
Seeley, David C.
Seely, David W.
Seely, Jeffrey B.
Segarra-Santiago, Raymond X.
Seideman, Charles H.
Seitz, John A. III
Sekel, Clifford Jr.
Selburg, Allan M.
Sellers, Link O.
Selner, Kurt J.
Semler, John
Sennewald, Robert W.
Sentell, Gerald D.
Serafini, Adrian J. Jr.
Serdich, Andrew P.
Sesto, Russell K.
Severn, Theodore
Sewell, James P.
Sexton, John H.
Sexton, Robin D.
Shadburn, Jeffrey S.
Shaffstall, Mark A.
Shannon, Christopher S.
Shapiro, Jeffrey M.
Sharp, Charles R.
Sharp, Michael A.
Sharpe, Thomas L.
Shattuck, Samuel H.
Shaughnessy, Robert E.
Shaw, Clarence R.
Shaw, John E.
Shaw, John M.
Shaw, Rollin R.
Shawn, Gregory L.
Shearin, James M. Jr.
Sheaves, William B.
Sheets, Robert E.
Shell, Leighton J.
Sheperd, Thomas E.
Sheridan, Cedric N.
Sheridan, Edward W.
Sheridan, Joseph C.
Sherkey, Mark B.
Sherman, Robert M.
Sherman, Scott M.
Sherrod, Steven G.
Sherwood, John T.
Sherwood, Richard W.
Shevis, Philip G.
Shields, Aaron R.
Shields, Christopher L.
Shields, Douglas E.
Shields, George S.
Shields, Hurley D.
Shinholt, Bryan E.
Shipman, Michael W.
Shirley, Tom L.
Shockley, Raul D.
Shoffner, Wilson A.
Shontz, Robert J. Jr.
Short, Charles E.
Shrewsberry, Harold F.
Shriner, Andrew C.
Shugart, David A.
Shumaker, Roger A.
Shumway, Andrew B.
Shupp, Charles B.
Shute, Benjamin C.
Shuto, James T.
Sickinger, Daniel L.
Sievers, Dean L.
Sifri, Nadeem S.
Sigler, Chad W.
Silber, Carl H. III
Siler, Jeffery B.
Siltman, Frank J.
Silvasy, Jeffrey M.
Simino, Joseph R.
Simmons, Randall V. Jr.
Simmons, Shawn C.
Simon, Michael
Simpson, Joseph P.
Simpson, Kenneth W.
Simpson, Larry L.
Sims, Eric Shelton
Sims, James A.
Sims, William R.
Sincere, Clyde J.
Singer, James H.
Singer, Michael L.
Singleton, George B.
Singleton, Raymond C.
Sinnreich, Richard Hart
Sipes, James T.
Sisco, Richard F.
Sisson, Rodney E.
Sitkiewitz, Scott T.
Sizelove, John M.
Skaggs, Douglas C.
Skeem, Stewart C.
Skelton, James R.
Skillern, Sean L.
Skilling, Jeffrey K.
Skinner, Charles R.
Skubi, Anthony J. III
Slater, Duane L.
Slaughter, Benjamin P.
Slee, Howard M.
Slezak, John W.
Slinsky, Ed
Sliwa, Steven A.
Slonina, John R.
Slucher, Albert T.
Small, Chris W.
Small, Thomas L.
Smith, Anthony M.
Smith, Arnold
Smith, Benjamin S.
Smith, Byron J.
Smith, C. Coburn
Smith, Carl W.
Smith, Charles J.
Smith, Christopher L.
Smith, Coniculus B.
Smith, Danny K.
Smith, Dennis C.
Smith, Eric T.
Smith, Eugene B.
Smith, Gary
Smith, Henley L.
Smith, Hyrum W.
Smith, James L.
Smith, James M. IV
Smith, James R.
Smith, James R. II
Smith, Jay Q.
Smith, Jeffrey D.
Smith, Jerry M.
Smith, Joey R.
Smith, John J.
Smith, John W.
Smith, Kent A.
Smith, Kevin J.
Smith, Loren K.
Smith, Malcolm G.
Smith, Martin J.

Smith, Melvin R.
Smith, Michael
Smith, Michael A.
Smith, Perry G.
Smith, Phillip J.
Smith, Reginald K.
Smith, Rickey E.
Smith, Robert B.
Smith, Robert L.
Smith, Ronald L.
Smith, Russell H.
Smith, Shad G.
Smith, Stephen F.
Smith, Steven E.
Smith, Taber S.
Smith, Thomas M.
Smith, Tony G.
Smith, Wayne E.
Smitherman, Jeffrey C.
Snapp, Patrick G.
Sniffin, Thomas D.
Snow, Richard A.
Snyder, John B.
Snyder, Pugh K.
Snyder, Robert D.
Soby, Charles S.
Soden, Roger L.
Soderberg, Bruce W.
Sofield, Michael J. Jr.
Solberg, William O.
Sole, William R.
Solomon, Joseph E.
Solomon, Norman E.
Sommerville, David L.
Sonneborn, Jay A.
Sossaman, Scott H.
Soto, Martin A
South, Mark J.
Southard, Floyd F.
Sowards, Jon K.
Spadaro, Anthony A.
Spalding, Sidney P.
Spann, Foy R.
Spann, Johnny M.
Sparkes, James A.
Sparrow, Joshua T.
Spates, Earnest H.
Spaulding, Benjamin M.
Spaulding, Logan T.
Speacht, Charles F. IV

Spears, Gregory W.
Speir, William B. Jr.
Speltz, Michael J.
Spencer, Keith L.
Spencer, Nicholas
Spiece, Donald C. Jr.
Spivey, Amos R. Jr.
Spivy, Berton E. Jr.
Spivy, Peter B.
Spohn, Edward A. III
Spohr, Robert E.
Sprackling, James T.
Spratlan, Walter D.
Sprayberry, Billy F.
Springer, Dean
Sprinkle, Joel M.
Spritzen, R. J.
Sprous, Gilbert A.
Spurdens, Alfred
Spurgin, Mark R.
Sridharan, N. S.
Stacy, Edwin W.
Stadler, Gerald P.
Stafford, Anthony T.
Staley, Terrance Eugene
Stallings, John A.
Stallings, Joseph L.
Stanfield, Timothy L.
Stanley, David L.
Stanovich, Mark E.
Stansel, Jeffery G.
Stapleton, Mark L.
Stark, Brent E.
Stark, Timothy R.
Stauffacher, Philip R.
Stawick, Joseph M.
Stebbins, Steven A.
Stecco, Johnny M.
Steck, Walter J.
Steele, Edward A.
Steele, John L.
Steele, Kevin P.
Steele, Ronald M.
Steers, Howard J. T.
Stefani, Richard N.
Stehlik, Jim
Stein, Thomas
Steiner, Christian W. Jr.
Steinke, Ralph R.
Stemple, Philip A.

Stephen, Garvin Keith
Stephens, Chester O. Jr.
Stephens, Darryl D.
Stephens, Elmer L.
Stephens, Jay D.
Stephens, John K.
Stephenson, Michael B.
Sterling, Joseph
Sterny, John M.
Sterrett, Lawrence D.
Stetson, Eric W.
Stevens, Barrett M.
Stevens, George E.
Stevens, Leigh E.
Stevens, Lynn H.
Stevenson, Douglas J.
Stevenson, J. L. (Buz)
Stewart, Danny
Stewart, David P.
Stewart, Dennis
Stewart, James C.
Stewart, James M.
Stewart, Jason P.
Stewart, Norman P.
Stewart, Shaun M.
Still, Douglas R.
Stimson, Theron F.
Stock, Paul A.
Stomprud, Larry C.
Stone, Ernest F.
Stone, Graham M.
Stone, Thomas R.
Stone, William B.
Storck, John C. III
Stothers, David E.
Stotts, Kenneth L.
Stowe, Lisa M.
Strachan, John M.
Strashinsky, Alex R. Jr.
Stratman, Frank
Stratman, Henry W.
Stratton, Bob
Straub, Matthew M.
Strawn, Marvin I.
Street, Arthur J.
Strickland, William C.
Strom, Charles P.
Stromp, Daniel P.
Strong, Brian L.
Strong, John L.

Strong, Robert L.
Stroumpos, George J.
Stuart, Lynn A.
Stucker, William E.
Stuckey, Rodridquez L.
Studdard, Samuel T.
Stull, Bill A.
Stumpf, James C.
Sturgess, Keith
Sublett, Carl L. Sr.
Suddreth, Steven L.
Sugerman, Leonard R.
Sughrue, Charles F.
Sullenger, Lee M.
Sullivan, John M. Jr.
Sullivan, Mike
Sullivan, Ricki Lynn
Sullivan, Timothy P.
Sullivan, William F.
Summa, Ronald J.
Summers, Glenn F.
Summers, Tom E.
Sumner, Gordon Jr.
Supko, Leonard M.
Sutherlin, Sharon L.
Sutton, Jason W.
Sutton, Lewis J.
Suver, Jeffrey A.
Swacina, Daniel J.
Swadley, Michael E.
Swanson, Harold M.
Swanson, Justin M.
Swartz, Leonard G.
Sweeney, Patrick J.
Sweetser, Nathan V.
Swenson, Gary G.
Swett, Frank M.
Swift, Richard J.
Swindell, David K.
Sydenham, Stanley R. Jr.
Sydenstricker, Paula A.
Syverson, Paul R. III
Szlenk, Marek
Tabor, David B.
Tadlock, Max E.
Tait, Curtis D.
Talbot, Charles E.
Talcott, Christopher P.
Talty, Thomas L.
Tamargo, James W.

Tannenbaum, David L.
Tanner, Carly A.
Tarantella, Thomas E.
Tarble, Burke A.
Tarrant, Joseph W.
Tassone, Vincent J.
Tastad, Jerry
Tatarek, Dale C.
Tatom, Michael B.
Tatroe, Richard L.
Tauber, Bernard L.
Tavuchis, Christopher A.
Taylor, Bernard
Taylor, Christopher P.
Taylor, Clarence L. Jr.
Taylor, Clark B.
Taylor, Dustin H.
Taylor, Jack S.
Taylor, Jeffrey A.
Taylor, John
Taylor, John D.
Taylor, Lawrence M.
Taylor, Lynn A.
Taylor, Patrick H.
Taylor, R. J.
Taylor, Ricky L.
Taylor, Robert E.
Taylor, Thomas D.
Taylor, Thomas E.
Teague, Kirby K.
Tegtmeier, Carl J.
Tempesta, Frank
Terrell, Eldon J.
Terrell, Ernest P. Jr.
Terrill, Jeffrey M.
Terry, Robert F.
Tesch, Gregory K.
Thacker, Harold A. Jr.
Thacker, Rodney A.
Thackston, Turner B.
Thallas, Ronnie S.
Thein, Scott E.
Theis, Joseph C.
Then, Kenneth M.
Theres, Gregory J.
Thewes, Richard
Thielemann, Chris J.
Thieman, Frank A.
Thomas, Barry W.
Thomas, Bobby Ray Jr.
Thomas, Christopher W.
Thomas, David F.
Thomas, George B.
Thomas, James A. Jr.
Thomas, Johnny F.
Thomas, Ricky L.
Thomas, Robert F.
Thomas, Tommy Jr.
Thompson, Billy R.
Thompson, David F.
Thompson, David S.
Thompson, Donald W.
Thompson, Ellwyn L.
Thompson, James B. Jr.
Thompson, John B.
Thompson, Jose M.
Thompson, Kevin S.
Thompson, Lloyd B.
Thompson, Robert F. Jr.
Thompson, Thomas C.
Thompson, Walter L.
Thomson, John C. III
Thorne, Brian E.
Thorne, Carroll R.
Thornhill, Terry L.
Thornton, Joe A.
Thrasher, Alan W.
Throwe, James H.
Thurlow, Jon M.
Thurmond, William D.
Tidwell, Mareld T.
Tierney, Charles B.
Tilley, James Dwight
Tillman, Mark E.
Timoshkov, Victor S.
Tincher, Ronald D.
Tinti, Wayne P.
Tipa, Ronald J.
Tippett, David A.
Tippetts, Isaac T.
Tipps, Jerry W.
Tipton, Jeffrey S.
Titus, Timothy N.
Tobey, Nelson W.
Todd, Jack
Todd, John R.
Todd, Michael A.
Todhunter, Eric J.
Tohda, Hideho
Tokar, Leonard G. Jr.
Toler, Alan G.
Toll, Art
Tolley, Stephen W.
Tolzmann, Mark A.
Tomany, Michael L.
Tomaschow, David W.
Tomasso, Richard P.
Tomcho, Timothy J.
Tomlin, William P.
Toney, Kenneth W.
Tooke, Shaun E.
Toomer, Jeffery K.
Tooson, Aubrey D.
Torrance, Steven M.
Torrance, Thomas G.
Torres, Christian E.
Torres, Ricardo R.
Torres-Lopez, Rene
Toth, George Jr.
Toupin, Robert E.
Tovey, Robert D.
Towers, John W.
Towers, Kristopher A.
Townsend, Ronald E.
Toy, Arthur
Tracy, Michael S.
Tracy, Tom J.
Tran, Pon V.
Tran, Steeve D.
Trantham, William E.
Travis, Stuart B.
Traylor, George T.
Traylor, Mark A.
Trefry, Richard G.
Tremblay, Kenneth R.
Trembly, Stuart
Tressler, Jacob W.
Trettin, Thomas P.
Trevino, Raymond A.
Trexler, Glenn H.
Trexler, Robert K.
Tribble, Dan R. II
Trickett, Gary F.
Triplett, Mark H.
Triplett, Michael W.
Trippe, Henry E.
Tritsch, Thomas M.
Tromblee, Mark L.
Trost, Jon L.
Trott, Earle C.
Troutman, Danny
Troutman, Timothy C.
Truax, Douglas L.
Truax, Sean R.
Trzcinski, Timothy R.
Tucker, Edwin J.
Tuemler, Aaron D.
Tuomey, Michael S.
Turinski, Drew A.
Turlington, John E.
Turner, Bradley
Turner, Curtis W.
Turner, Earl
Turner, Edward M.
Turner, Frank D. III
Turner, James L.
Turner, J. Scott
Turner, Joseph W.
Turner, Michael A.
Turner, Mike B.
Turner, Philip L. III
Turner, Ralph R.
Turner, Rondal L.
Turner, Troy J.
Turner, Wendell H.
Turner, William B. Jr.
Tuttle, Richard D.
Tutton, Stephanie J.
Tyler, Bryan A.
Tyler, Terry J.
Tyson, Dirk W.
Tyson, John M.
Uberti, John
Ukperaj, Salih
Ulitzsch, Jason W.
Ullian, Richard P.
Ullom, Steven C.
Umphrey, Fred E.
Underwood, Johnny W.
Unger, F. T.
Unsal, Bulent
Unterseher, James E.
Utley, Lance D.
Vacchi, David T.
Vail, Richard H.
Valcourt, David P.
Valentine, John M.
Valenzuela, Alfred A.
Vallario, Richard A.
Vallentine, John M.

Vanak, Paul A.
VanAlstine, Aaron J.
VanBuren, Stewart J.
VanCleave, Jeff T.
Vandal, Thomas S.
Vandenbosch, Eric J.
Van Dyke, Jeffrey B.
VanFleet, Townsend A.
Vangjel, Peter M.
VanKeuren, Edwin J.
Van Pelt, Harvey A. Jr.
VanPool, Jack L.
Van Sant, John E.
VanTilburg, Wayne
van Vliet, A. L. H.
Vanwagner, Dennis W.
Varner, James E.
Varney, Ernest E.
Vasquez, Manuel D.
Vasso, Amery E.
Vaughan, James D.
Vaupel, Leon D.
Vavra, Alvin E.
Vazquez-Roman, Cesar O.
Velasco, Anthony S.
Verge, Clinton D.
Verzola, Eric D.
Vessell, Charles
Vick, Michael D.
Vickrey, James D.
Victorino, James P.
Villandre, Alan J.
Villandre, Michael R.
Villegas, Francisco Nieto
Vilven, John R.
Vinson, David S.
Vinyard, Sherman L.
Visbal, Robert M.
Vitacolonna, Edward E.
Vital, Marc W.
Vizzard, James W.
Vockery, William N.
Vogel, John P.
Vogt, William C.
Vokac, Robert B.
Vokac, Robert H.
Vold, Jeffrey D.
Vollintine, James R.
Vonderschmidt, William W.
vonGerichten, Theodore C.
von Recum, Derik F.
vonRoeder, Steven S.
Vordermark, Jeffrey D.
Vozzo, Martin L.
Vuillemot, Andrew J.
Vuono, Carl E.
Vuono, Jeffrey E.
Vuono, Timothy A.
Wackerman, Damian C.
Wade, Chris A.
Wade, M. J.
Waggoner, Freddie R.
Wagner, Alan R.
Wagner, Robert A.
Wakeman, James H.
Walcott, Robert
Waldeck, James J. III
Waldorf, Michael
Waldrep, Matthew E.
Waldrop, Johnny L.
Walker, James M.
Walker, Merle T.
Walker, Michael L.
Walker, Richard E. III
Walker, Tommie L.
Walker, Walter M. Jr.
Wall, Timothy C.
Wallace, Edward T.
Wallace, John R.
Wallace, Raymond E.
Wallace, Ryan D.
Wallace, Vincent M.
Wallace, Wilbur W.
Waller, Christopher
Waller, Thomas G. Jr.
Wallestad, David K.
Wallington, Clinton J. III
Wallis, Jonathan A.
Walls, Louis J.
Walsh, Glenn A.
Walsh, Kenneth J.
Walsh, Michael T.
Walsh, Thomas L.
Walters, Paul R.
Walters, Robert P.
Waltman, Brian F.
Wampler, Bryan W.
Wang, Peter Y.
Wanner, Peter P.
Wanovich, James A.
Ward, Michael A.
Ward, Stephen C.
Ward, Windsor E.
Warden, Cumberland A.
Warden, John M.
Ware, Bobby D.
Waring, James M.
Warner, Charles H.
Warner, Christopher L.
Warner, Christopher R.
Warner, Monroe P.
Warnock, Tracy T.
Warnock, William H.
Warren, Gary R.
Warren, Joe W.
Warren, Tarn D.
Warrington, Edgar L.
Warshawsky, Christopher
Washington, Carlton L.
Washington, Michael F.
Wason, John D.
Waters, Douglas T.
Waters, Jason C.
Watkins, J. D.
Watkins, Stanley H.
Watson, John D. Jr.
Watson, Robert E.
Watson, Robert P.
Watson, Thomas E. Jr.
Watts, J. C. Jr.
Watts, Walter J.
Wax, Richard R.
Wayland, Gary M.
Weafer, Thomas W.
Weatherford, Darrell J.
Weatherspoon, Lynn A.
Weatherstone, Andrew J.
Weaver, Jeffrey B.
Weaver, Michael R.
Weaver, Vince C.
Webb, Aaron
Webb, Charles W.
Webber, Craig A.
Weber, Eric G.
Weber, Joseph F.
Weber, L. Lodge
Weber, Michael A.
Webley, J. R.
Webley, Steve
Webster, David L.
Wecas, Ron N.
Wedvick, Dwyer Q.
Weeks, William P.
Weese, Matthew G.
Wegenhoft, John F.
Weick, David S.
Weiford, Thomas E.
Weiser, James A.
Weiss, Thomas J.
Weissenberg, Scott D.
Weitzel, Robert J.
Weko, Charles W. III
Weller, Gary P.
Wellons, Dave
Wells, Arch W.
Wells, Arthur D.
Wells, Benjamin D.
Wells, Lewis C.
Welty, Kenneth E.
Wendell, Jason A.
Wendland, Christopher W.
Wenger, Gregory G.
Wertzler, Brian F.
Wesch, Peter H.
Wesley, Charles M.
Wesley, Michael F.
Wessling, James A.
Wesson, Kennedy L.
West, Aaron S.
West, Billy J.
West, Fred D.
West, Jeff N.
West, Richard L.
West, Stephen K.
West, Steven R.
Westbrook, Walton L.
Westcott, James F.
Westley, Scott A.
Wetterstroem, Robert S.
Weyrauch, Paul C.
Whalen, Donald P.
Whalen, Martin J.
Wharton, Bobbie F.
Wheeler, Larry L.
Whitaker, Edward S.
Whitbeck, Paul M.
White, Aubrey
White, Barney D.
White, Charles S.
White, Dan A.

White, David C.
White, James R.
White, James S.
White, Kevin D.
White, Kirk R.
White, Samuel R.
Whiteside, Daniel L.
Whiteside, Lee J.
Whitlock, William H.
Whitmire, John W.
Whitney, Dallas C.
Whitsett, Charles R.
Whittemore, Ryan H.
Why, Robert A.
Wick, Gregory J.
Wicker, Patrick S.
Wickliffe, Paul T.
Wiegand, Mark E.
Wiegers, Robert E.
Wielinski, Todd D.
Wieskamp, J. F. W.
Wiles, Richard I.
Wiley, Curtis D.
Wilhelms, John C.
Wilkerson, Terry G.
Wilkes, Hershel B.
Wilkinson, John R.
Willard, James E.
Willebrand, Guy E.
Willey, William T.
Williams, Charles M.
Williams, Christian E.
Williams, Daniel T.
Williams, Darryl A.
Williams, David L.
Williams, Donald E.
Williams, Glenn S.
Williams, Henry L.
Williams, James J.
Williams, John C.
Williams, John C. Jr.
Williams, John L.
Williams, John M. Jr.
Williams, Randall C. Jr.
Williams, Ryan K.
Williams, Terrence P.
Williams, Thomas M.
Williams, Vernon J.
Williams, Wendell T.
Williamson, Mark B.
Williamson, Richard W.
Willingham, Dwight M.
Willingham, Joe D.
Willis, James L.
Willis, John B.
Willis, Michael S.
Willis, William M.
Wills, Daniel
Willson, Thomas F.
Wilson, David S.
Wilson, Donald W. Jr.
Wilson, Jerre W.
Wilson, Larry E.
Wilson, Leigh R.
Wilson, Mark A.
Wilson, Patrick D.
Wilson, Sean P.
Wilson, Thomas G.
Wilson, Timothy I.
Winchell, Ronald E.
Wing, Charles J.
Wing, Leslie G.
Wingenbach, Karl E.
Wingfield, George D.
Winkelman, Perrin D.
Winkelmann, Joseph C. Jr.
Winter, Brian D.
Winters, Jeffrey S.
Winton, Douglas W.
Wise, Daniel V.
Wise, George R.
Witherspoon, Willie R.
Witschonke, Carl F.
Witt, Jeffrey S.
Woeltjen, Donald A. Jr.
Wolfe, Jack L.
Wolfe, Rodney D.
Wollen, Evan H.
Wollenhaupt, Charles L.
Wolters, John B.
Womack, John T.
Wong, George
Wong, Theodore L.
Wood, Eddie
Wood, J. W.
Wood, Jean-Yves R.
Wood, Jerry L.
Wood, Joe A.
Wood, John
Wood, John D.
Wood, John R.
Wood, Michael R.
Wood, Thomas E.
Wood, William J.
Woodford, Jason A.
Woodhead, Matthew A.
Woodlock, Richard
Woodruff, Dwight
Woodruff, John T.
Woods, James T.
Woods, John C.
Woods, Kevin B.
Woods, William H.
Woolcock, Bret T.
Wooldridge, Robert T.
Woolery, Robert H.
Wooley, James H.
Woolley, James W.
Wooster, Howard N.
Workman, Steven T.
Worley, Joe R.
Worley, Mark A.
Worley, Seward C.
Wortman, John F. Jr.
Wotring, John J. IV
Wreford, David V.
Wright, Frank H. IV
Wright, James L.
Wright, Kathy J.
Wright, Mark E.
Wright, Richard W. Jr.
Wright, Robert A.
Wright, Stewart M.
Wruck, Erich-Oskar
Wu, Victor P.
Wurman, James W.
Wyckoff, Frank M.
Wyman, Scott L.
Wynarsky, Andrew E.
Wynns, D. Bryan
Yamaguchi, Koji
Yamasaka, Yasuaki
Yancey, Andrew C.
Yang, Brian S.
Yates, Dennis W.
Yee, Jay W.
Yenchesky, James H.
Yi, Kenston K.
Yingling, John A.
Yockey, James D.
Yoder, Don
Yohe, Richard W.
York, Ronald S.
Young, Earl W. Jr.
Young, John L. III
Young, Morris M.
Young, Paige Allen
Young, Richard A.
Young, Robert F.
Young, Stephen A.
Young, Walter F.
Younger, Guy C.
Yowell, Michael L.
Yuengert, Louis G.
Yuschak, Michael
Zabecki, David T.
Zachar, Mihaly A.
Zachariasen, Craig Z.
Zacks, Y. Joseph
Zahn, Matthew A.
Zak, Leo P.
Zaramba, Glenn R.
Zavecz, Les W.
Zeeman, Eric W.
Zeller, Andrew S.
Zemp, William H.
Zenker, Ernest G.
Zepp, William T.
Zielenski, Peter T.
Zielinski, Peter J.
Zieska, Kenneth W. Jr.
Zimm, Alan D.
Zimmerman, Brian E.
Zimmerman, John B.
Zimmerman, Matthew H.
Zimmerman, Patrick J.
Zimmerman, Raymond C.
Zimmermann, Wayne L.
Zimni, Werner
Zino, Joseph
Zipp, Bernard F.
Ziter, Mark A.
Zoglman, Robert R.
Zuchelli, Kevin M.
Zuna, Luis F.
Zurschmit, Scott W.
Zwirecki, Joseph J.
Zybura, Martin A.
Zyla, Robert C.

Index